Empowering the
Great Energy Transition

Empowering the
Great Energy Transition

Policy for a Low-Carbon Future

Scott Victor Valentine,
Marilyn A. Brown, and
Benjamin K. Sovacool

Columbia University Press

New York

Columbia University Press
Publishers Since 1893
New York   Chichester, West Sussex
cup.columbia.edu

Library of Congress Cataloging-in-Publication Data
Names: Valentine, Scott V., author. | Brown, Marilyn A., author. |
    Sovacool, Benjamin K., author.
Title: Empowering the great energy transition : policy for a low-carbon future /
    Scott Victor Valentine, Marilyn A. Brown, and Benjamin K. Sovacool.
Description: New York : Columbia University Press, 2019. | Includes bibliographical
    references and index.
Identifiers: LCCN 2019007061 (print) | LCCN 2019980756 (e-book) |
    ISBN 9780231185967 (cloth ; acid-free paper) | ISBN 9780231546423 (e-book)
Subjects: LCSH: Energy policy—United States. | Clean energy industries—
    United States. | Renewable energy sources. | Climate change mitigation—
    Political aspects.
Classification: LCC HD9502.U52 V35 2019 (print) | LCC HD9502.U52 (e-book) |
    DDC 333.790973—dc23
LC record available at https://lccn.loc.gov/2019007061 LC e-book record available at
    https://lccn.loc.gov/2019980756

Columbia University Press books are printed on permanent
and durable acid-free paper.

Printed in the United States of America

Cover image: © plus49/Construction Photography/Avalon/Getty Images
Cover design: Lisa Hamm

Scott Valentine dedicates this work to Rebecca and Elle, my stars who help guide me. In loving memory of Ellery and Victor.

Marilyn Brown wants Frank and Katie Southworth to know how much their intellectual and moral support has meant.

Benjamin Sovacool wishes to thank his family as well for their continued support, but also four important mentors who have helped him grow personally and intellectually at critical moments in time: Brent Brossmann at John Carroll University, George Ziegelmueller at Wayne State University, Richard Hirsh at Virginia Tech, and, indeed, Marilyn Brown.

Contents

Preface

Cities, companies, and citizens across the globe are strengthening their commitment to sustainable practices and are considering alternatives to carbon-intensive and resource-abusive development pathways. Hotter summers, rising sea levels, forest fires, and climate extremes are beginning to persuade even recalcitrant onlookers that something is amiss. Success requires a transition in how we produce and consume energy, because the global reliance on fossil fuels has been one of the biggest—if not THE biggest—cause of pollution and environmental degradation in the world.

Fortuitously, low-carbon renewable energy systems have advanced remarkably and are more affordable than ever, and radical solutions for improving energy efficiency are now cost-competitive. But there are forces that resist change; their influence has made most governments unwilling to put a premium on the price of fossil fuels to reflect the human health and environmental damages they are causing. As a result, new policies and new business models are needed to mobilize and accelerate the clean energy transition.

Empowering the Great Energy Transition argues that a transition away from carbon-intensive energy sources is inevitable. It offers ways to expedite the transition, buoyed by a decade of technology breakthroughs.

At the same time, the book places issues of energy and climate justice front and center. A focus on equity is particularly important when public resources are being deployed or when vulnerable groups are affected. *Empowering the Great Energy Transition* takes up these themes and more.

Readers will be taken on a journey through the energy world as it is today; they will explore the numerous barriers to change and learn about the many developments that give us reason for optimism. With just a few adjustments, we can set humanity on a course to a new energy future that empowers entrepreneurs and communities, inspires innovation, and mitigates the damages caused by conventional energy technologies that have served their purpose. We describe the achievable steps that citizens, organizational leaders, and policymakers can take to put their commitments to sustainability into practice. *Empowering the Great Energy Transition* shows that, with just a few adjustments, we can set humanity on a course that motivates and supports entrepreneurs and communities that seek a cleaner energy future.

In sum, this book:

- Is written by authors who together have produced over thirty books on energy and who are recognized experts in the field.
- Begins with a strong and surprising assertion, supported by ten key market trends, that indicates that a transition away from carbon-intensive energy technologies is unavoidable.
- Clarifies where the uncertainties of climate science lie and why these uncertainties impede market evolution.
- Provides a politically infused assessment of the obstacles that are slowing a low-carbon energy transition.
- Concludes with specific prescriptions to consumers, organizational leaders, and policymakers on how they can make a difference.

Acknowledgments

All three authors want to thank the editors and reviewers at Columbia University Press for their excellent suggestions for improvement. Support for this book was provided by the Brook Byers Institute of Sustainable Systems, the School of Public Policy and several engineering and science programs at the Georgia Institute of Technology that funded a workshop held in Atlanta in July 2017 to discuss interdisciplinary aspects of the global energy transition. That workshop enabled us to advance our concepts of technical and societal transformations with an audience that was deeply engaged in developing technology solutions to clean energy and climate change challenges.

Empowering the
Great Energy Transition

1 | The Great Energy Transition

Anyone who has experienced the joys of child-raising has had an "I knew that was going to happen" moment—a phrase often uttered as a cup half-filled with some sort of stain-producing agent is launched off a table by a pudgy-armed baby. The agent bounces miraculously off the toe of your slipper before skittering across the floor to the foot of your new cloth designer bag. Then the lid of the cup that was holding the liquid comes unhinged, unleashing a torrent onto the kitchen floor and launching your new bag afloat on a sea of stain. But of course, you actually didn't know *that* was going to happen. You only suspected that something *like* that *might* happen. The likelihood of the cup departing on a journey of this type was, in your mind, perilously high. But the specific damage incurred was not preordained; it was merely a manifestation of Murphy's Law. In short, this was a case of uncertain certainty, wherein general outcomes can be vaguely predicted, but not with specificity.

Uncertain certainty describes the energy sector today. There are clear indications that the energy sector—defined as the sociotechnical systems that supply and distribute energy services such as heat, mobility, hydraulics, and steam—is transitioning away from carbon-based energy resources. Yet these trends do not impart specific insight into what the sector will transition to or how this transition will unfold. Predicting an endgame is of value for those seeking to predict future energy costs, technology-specific market trends, or geopolitical developments associated with energy transitions. But how can one do this? We know that the cup is about to come off the table, but where will it land?

Before we consider these conundrums in greater depth, why we are so sure that a transition is under way warrants further attention. After all, if a comprehensive transition is not afoot, there is little need for reading on.

TEN CURRENT TRENDS POINTING TO A TRANSITION

If we were to look retrospectively over energy trends since the dawn of the new millennium, it would not be an encouraging landscape in regard to CO_2 emissions. As table 1.1 indicates, despite a global imperative to transition out of carbon-based energy systems to reduce the social, economic, and ecological damages of climate change, energy generated from fossil fuels actually increased slightly from 80.3 percent to 80.7 percent

TABLE 1.1 Global primary energy supply by source

Energy source	2000		2017	
	mtoe*	Market share (%)	mtoe*	Market share (%)
Coal	2,308	23	3,750	26.8
Oil	3,665	36.6	4,435	31.7
Gas	2,071	20.7	3,107	22.2
Nuclear	675	6.7	688	4.9
Hydro	225	2.2	353	2.5
Modern bioenergy	377	3.8	727	5.2
Solid biomass	646	6.4	658	4.7
Other renewables	60	0.6	254	1.8
Total	10,027		13,972	

Source: International Energy Agency (IEA), *World Energy Outlook 2018* (Paris: IEA), 38.

*mtoe = million tonnes of oil equivalent.

market share between 2000 and 2017. To compound the problem, over this seventeen-year period, total global consumption of energy increased by 39 percent (10,027 mtoe to 13,972 mtoe), resulting in an undesirable increase in annual CO_2 emissions from 23.1 gigatonnes (Gt) to 32.6 Gt (mtoe = million tonnes of oil equivalent).[1]

Over this same seventeen-year period, some influential world leaders gave their take on the phlegmatic transition, placing the blame on the expense of renewable energy technologies. George W. Bush, during his U.S. presidency, explained the rationale for refusing to ratify the Kyoto Protocol and its transitionary targets in the following way: "Kyoto would have wrecked our economy. I couldn't in good faith have signed Kyoto."[2] The implication was that a renewable energy transition would undermine the U.S. competitive advantage. Former Canadian prime minister Stephen Harper, who as far back as 2002 labeled the Kyoto Protocol a "socialist scheme to suck money out of wealth-producing nations,"[3] disengaged Canada from the Kyoto Protocol within a year of coming to power in 2006. Former Chinese premier Wen Jiabao framed Chinese policy by warning that "action on climate change must be taken within the framework of sustainable development and should in no way compromise the efforts of developing countries to get rid of poverty."[4] In December 2013, former Australian prime minister Tony Abbott stated, "We have to accept that in the changed circumstances of today, the renewable energy target is causing pretty significant price pressure in the system and we ought to be an affordable energy superpower . . . cheap energy ought to be one of our comparative advantages . . . what we will be looking at is what we need to do to get power prices down significantly."[5]

In 2017, President-elect Donald Trump signaled a desire to place support for fossil fuels at the center of his administration's energy agenda by appointing a pro–fossil fuel politician, Scott Pruitt, to head the Environmental Protection Agency; Rick Perry, a Texan who has refused to accept anthropogenic activities as drivers of climate change, to head the Energy Department; and former ExxonMobil CEO Rex Tillerson to head the State Department. Although Pruitt has since resigned amid a misuse of funds scandal and Tillerson has been replaced after falling out of favor with Trump, the U.S. president clearly feels that fossil fuels serve economic growth aspirations.

Due to extensive media conditioning, the average person on the street may not even bat an eye when exposed to such statements and developments. For decades, climate change denial and inflated estimates about the high cost of renewables have pervaded national energy policy rhetoric in many nations with the staying power of odious cologne. So there must be something to all of this, right? Renewable energy is prohibitively expensive, and therefore its diffusion will stagger slowly forward fueled by government subsidies until commercial viability finally arrives.

Evidence suggests that this perspective is incorrect. A virtually unassailable body of evidence indicates that strong forces favoring renewable fuels have already materialized and are altering market dynamics in favor of a transition to renewables. This chapter will examine ten such drivers:

1. Growing evidence of declining fossil fuel stocks and rising prices
2. Capricious fluctuation patterns of fossil fuel prices
3. The strategic need to diversify
4. Political instability and conflict due to fossil fuels
5. Improved understanding of fossil fuel health and environmental costs
6. Sobering evidence of climate change impacts
7. The contested politics of nuclear power
8. Innovations in performance and cost within the renewable energy sector
9. The rise of government and market support for renewable energy
10. First-mover advantages amid a new energy boom

Trend 1. Growing Evidence of Declining Fossil Fuel Stocks and Rising Prices

Humans have been burning through fossil fuel reserves as if there were no limits. Yet the finite nature of fossil fuel reserves suggests one inevitable consequence: stocks will progressively diminish. As table 1.2 illustrates, even if humanity stopped fossil fuel discovery efforts to meet burgeoning demand (i.e., satisfying demand growth through renewable energy), proven reserves in oil would last only 51 years and proven natural gas reserves would be

TABLE 1.2 Life expectancy of proven fossil fuel resources, 2016

Energy source	Proven reserves	Current production	Reserve-to-production ratio (years)		
			0% annual production growth rate	1.0% annual production growth rate	2.3% annual production growth rate
Oil	1,698 billion barrels	33.5 billion barrels	51	41	33
Natural gas	187 trillion cbm*	3.5 trillion cbm	53	43	34
Coal	891,531 million tonnes	7,820 million tonnes	114	76	54

Source: Reserve and production data from *BP Statistical Review of World Energy*, June 2017 (authors' calculations).
*cbm = cubic meters.

used up in 53 years. The International Energy Agency estimates that global energy demand will grow by 1 percent under its new policy scenario up until at least 2040. This contrasts with a 1.4 percent annual growth under current policies. It further estimates that fossil fuels will continue to deliver the bulk of our energy.[6] As table 1.2 ominously demonstrates, if fossil fuels experienced a 1 percent annual growth in demand, reserve-to-production ratios would decline to 41, 43, and 76 years, respectively, for oil, gas, and coal. Moreover, as the far-right column of table 1.2 illustrates, if fossil fuel energy production expanded at the same annual rate recorded between 2000 and 2014 (2.3 percent), we would face the full depletion of current proven reserves within the next three to four decades.

The analysis presented in table 1.2 is disingenuous in that it omits an important fact: to date, fossil fuel companies have been expanding proven reserves through exploration and discovery efforts, and innovations also continue in other domains such as advanced oil recovery, imaging, drilling, and transport. Therefore, a counterargument can and should be made that proven reserves represent only a portion of the reserves that fossil fuel firms will uncover in the future.

TABLE 1.3 Expansion of proven fossil fuel reserves, 1995–2017

Fuel stock	1995 proven reserves	2017 proven reserves	Change	% change
Oil	1,126 billion barrels	1,696 billion barrels	570 billion barrels	51
Natural gas	120 trillion cbm*	193 trillion cbm	73 trillion cbm	61
Coal	1,031,610 million tonnes	1,035,012 million tonnes	3,402 million tonnes	0.3

Source: Reserve and production data from *BP Statistical Review of World Energy*, June 2018 (authors' calculations). *cbm = cubic meters.

There are two caveats to the cornucopian view that such an argument conveys. First, as table 1.3 illustrates, substantial discovery efforts have just barely been able to stave off the depletion of coal reserves. At the end of 2015, proven coal reserves amounted to 891,531 million tonnes.[7] Since then, massive discovery efforts have been under way to unearth more coal to keep up with amplified demand. However, these efforts have just been able to maintain steady reserves, suggesting that the down button on this particular commodity elevator has been pushed. Consequently, developing policy around an assumption that unproven reserves might be sufficient is akin to concluding that since we do not know when we are going to die, we should plan to live forever.

Second, the expansion of oil and gas reserves has so far kept pace with the expansion of demand, but it has come at a price, literally. Discovery efforts are increasingly focused on more expensive sites (deep sea wells, Arctic locations, etc.). As we move forward, adding to reserves will come at a much higher cost to the end-consumer.[8] There will be exceptions, as the shale gas boom in the United States illustrated, but just like the shale gas bonanza, these exceptional finds will be short-lived. The norm going forward will entail higher discovery costs feeding into higher end-consumer pricing. Bethany McClean goes further by forecasting that the next financial crisis lurks underground, fueled by debt and years of easy credit made available to the U.S. fracking industry.[9]

Energy economics tells us that, barring any significant reduction in demand for fossil fuels, dwindling stocks will eventually engender price inflation, whether that is because of scarcity or the higher cost of discovery. Indeed, there is ample evidence that supply expansion efforts are no longer as successful in curtailing price inflation as they have been in the past.[10] The figures that follow highlight the deteriorating economics of fossil fuels.

Bituminous coal is higher-grade coal (compared to lignite), often used along with sub-bituminous coal in thermal power plants. Up until the 1970s, the price per short ton of bituminous coal in the United States was consistently under US$7. As a consequence of the oil crises of the 1970s, a global flight from oil-fired power generation to coal-fired power generation caused a demand-driven escalation of coal prices. As is the case with all fossil fuels, as commodity prices rise, discovery efforts intensify. In the 1980s these efforts proved successful, and the price of a short ton of bituminous coal settled into a US$20 to US$30 range,[11] higher but still economically palatable. Since 2004, a global surge in electricity demand—led by developing nations such as China and India—has pushed coal prices even higher. Yet once again suppliers have responded with renewed discovery efforts. Consequently, from a peak in 2010, the price of bituminous coal averaged US$55.60 per short ton in 2017,[12] culminating in an average coal-fired electricity generation price in the United States of US$2.06 per million Btu (MBtu).[13] Going forward, the U.S. Energy Information Administration conservatively forecasts that the cost of coal will rise to US$2.56/MBtu by 2050.[14] Nevertheless, this is a further 25 percent cost increase in a coal-rich nation.

A trend of upwardly ratcheting prices should be apparent in this narrative; amplified demand pushes prices higher, and suppliers respond by escalating discovery efforts, which pull prices back a bit, only to spur additional demand increases. Coal prices are rising as economic theory predicts, sporting an inflated price level that is five to six times what it was in the 1970s. Figure 1.1 illustrates the impact that the escalating price of coal has had on power plant operation costs.[15] This is one roller coaster that will not return to its embarkation platform.

Turning to the cost of natural gas, figure 1.2 presents a picture of what has been occurring in the U.S. Henry Hub natural gas market,[16] which can be considered a conservative proxy for global price trends (Japanese, German, and United Kingdom gas market prices are much higher[17]). Over the past fifteen years, natural gas prices have been very volatile. From a base of about US$2/MBtu in 2002, the cost of natural gas in the United States

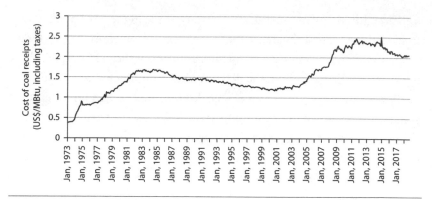

FIGURE 1.1 U.S. power plant coal costs, 1973–2018, nominal US$ (U.S. Energy Information Administration, 2018, https://www.eia.gov/totalenergy.)

spiked sharply, with prices topping US$18/MBtu in 2003, US$15 in 2006, and US$13 in 2009, all the while ratcheting up higher during nonpeak times as well. This trend mirrors the cycle described for coal: demand increases, driving prices higher; then, suppliers intensify discovery efforts, causing prices to fall back a bit before prices once again begin an ascent as demand increases. All of this is tied to economic growth. Between 2002 and 2009, global GDP nearly doubled, from US$34.4 trillion to US$63 trillion.

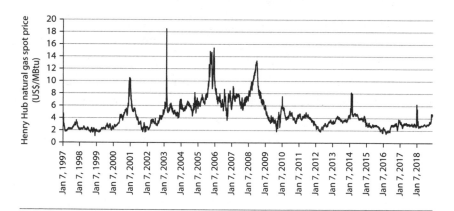

FIGURE 1.2 Henry Hub natural gas price trend, 1997–2018, nominal US$ (U.S. Energy Information Administration, *Natural Gas*, accessed December 14, 2018, https://www.eia.gov/dnav/ng/hist/rngwhhdD.htm.)

Since 2012, however, the price of natural gas in the United States has fallen back to nearly premillennium levels, and politicians are now mopping their brows and congratulating themselves for riding out the storm. The trouble is, this is just the beginning of storm season in the natural gas market. Low natural gas prices in the United States are largely due to a massive supply expansion catalyzed by shale gas discoveries. According to the U.S. Energy Information Administration (EIA), proven gas reserves in the United States grew from 60,644 billion cubic feet (bcf) at the end of 2009 to 199,684 bcf by the end of 2014.[18] That's the good news. The bad news is that natural gas production in the United States topped 13,000 bcf in 2014 and blossomed to 27,000 bcf in 2015.[19] IEA forecasts that U.S. gas production will expand by nearly 40 percent to 37,400 bcf in 2040, exceeding the growth rate of the Soviet Union's record gas expansion in 1974–1989.[20] But prices will rise as U.S. producers are forced to move away from "sweet spots" to exploit less productive zones. As a result, low natural gas prices will be short-lived. When reserves are exhausted, economic theory suggests that both gas and coal prices will rise, as will the angst of energy policymakers, who were banking on an extended glut of cheap natural gas. The EIA aligns with this view, forecasting that the cost of natural gas will rise 55 percent by 2050 relative to the 2018 average, from US$3.62/MBtu to US$5.60/MBtu.[21]

The history of oil prices tells a story of euphoria and woe since 1973. The drivers behind this extreme market turbulence merit review to better understand the evolution of oil market dynamics. The ratcheting up of prices in the 1970s (US$11/barrel to US$55/barrel to US$107/barrel) tells the economic story associated with the OPEC-centered supply disruptions. In 1973, the OPEC nations colluded to stop sales of oil to the United States in retaliation for America's involvement in the Yom Kippur War. In 1979, a revolution in Iran led to a subsequent war between Iran and Iraq, crippling oil production in both nations. On the heels of these two supply disruptions, nations around the world began to shift away from oil-fired electricity generation, and a global hunt for oil reserves outside the Middle East commenced. Demand curtailment and supply expansion precipitated a gradual decline in oil prices between 1980 and 2001. Nevertheless, in inflation-adjusted dollars, oil prices in 2000 were hovering around the US$40/barrel level, a significant increase from the $10 to $20 range experienced between 1930 and 1970.[22]

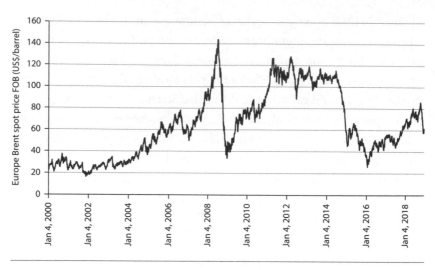

FIGURE 1.3 Crude oil prices, 2000–2018, nominal US$ (U.S. Energy Information Administration, 2018.)

Since 2000, though, stressors have changed. To better highlight the nuances of market dynamics in this period, figure 1.3 displays Brent oil prices solely during that period. First, there were the 9/11 terrorist attacks in the United States. The subsequent U.S. military operations in the Gulf region led to a curtailment of supply, causing oil prices to take flight from a base of around US$23 per barrel. While this modern-day version of the Vietnam War was being carried out, the global economy was surging, propelled by huge economic leaps forward by China and India. This trend precipitated higher demand for oil, amplifying upward pressure on oil prices. As energy security expert Daniel Yergin has observed, this was the first time in the history of fossil fuels that demand-side pressures became the dominant influence on fossil fuel energy prices.[23]

As has been the case for both coal and gas, high oil prices drive amplified discovery efforts. From 2008 on, oil production expanded in the United States at an unprecedented rate.[24] Supply expansion plus a global economic recession caused a 40 percent freefall of oil prices in 2009. Yet, like a resilient boxer, oil prices gradually began a comeback in 2010. Once again, economic expansion in China, India, and other emerging economies drove demand-side increases that outstripped the pace of supply expansion. By 2011, oil had topped new heights, selling in excess of US$110/barrel.

There is evidence that the price decline since 2012 has been a result of massive shale gas discoveries in the United States that continue to suppress demand for renewable fuels.[25] Moreover, the global economy continues to stagnate, further attenuating demand-side pressures. Yet, with automotive sales in China alone now exceeding 21 million cars per year and growing at a 7 percent annual rate,[26] a wise bet would be that demand for oil will continue to grow and, consequently, future oil price levels will rise. As an indication that the oil elevator is once again heading to the penthouse, it merits noting that the Brent Crude spot price on December 7, 2018, was US$61/barrel. On the same day, two years earlier, the same barrel of oil was selling for US$52, and three years earlier for US$40.[27] This escalation of oil prices is sure to set off another round of elevated discovery efforts, which might temper the pace of the upward trajectory, but oil prices can be expected to continue to rise. The button for the penthouse was pushed a long time ago, and there is not much we can do about it now—unless of course we wean ourselves from dependence on this finite resource.

The International Energy Agency (IEA) concurs with this forecast of increasing oil prices. As global economic activity progresses, barring any sudden transition out of carbon-based energy systems, fossil fuel energy prices are expected to climb. Table 1.4 outlines fossil fuel price forecasts under three scenarios. The "new policies" scenario assumes that all current policies plus current government pledges for CO_2 emission reductions are actualized (an optimistic business-as-usual scenario). The "current policies" scenario assumes that only committed policies are met (a business-as-usual scenario). The "sustainable development" scenario assumes that nations fully meet their Paris Agreement promises to keep climate change to under 2°C (a very optimistic scenario). Under every scenario, the prices of oil and natural gas are expected to rise significantly by 2040. The price of coal is expected to rise in every scenario—even in the "sustainable development" scenario, which is based on a mass exodus from fossil fuels. Commodity prices from 2016 and 2017 have been included in this table to convey a further point of concern: given the volatility in prices from 2000 to 2017 (and even between 2016 and 2017),

TABLE 1.4 IEA fossil fuel import price estimates

Real terms (US$2017)	Actual prices				New policies				Current policies		Sustainable development	
	2000	2010	2016	2017	2025	2030	2035	2040	2025	2040	2025	2040
IEA crude oil (US$/barrel)*	39	88	41	52	88	96	105	112	101	137	74	64
Natural gas (US$/MBtu)												
United States	6.0	4.9	2.5	3.0	3.3	3.8	4.3	4.9	3.4	5.3	3.3	3.6
European Union	3.9	8.4	4.9	5.8	7.8	8.2	8.6	9.0	7.9	9.4	7.5	7.7
China	3.6	7.5	5.8	6.5	9.2	9.4	9.5	9.8	9.3	10.2	8.3	8.5
Japan	6.6	12.3	7.0	8.1	9.8	10.0	10.0	10.1	9.9	10.5	9.0	8.8
Steam coal (US$/tonne)												
United States	38	64	49	60	63	63	64	64	64	69	58	56
European Union	47	103	63	85	80	83	84	85	84	98	69	66
Coastal China	35	130	NA	102	91	93	94	94	95	106	81	79
Japan	45	120	72	95	85	88	89	90	89	105	74	70

Source: International Energy Agency (IEA), *World Energy Outlook 2018*, https://www.iea.org/weo2018/.

*The IEA crude oil price is a weighted average import price among IEA member countries.

one wonders whether these relatively stable projections between 2025 and 2040 will indeed be indicative of reality.

Trend 2. Capricious Fluctuation Patterns of Fossil Fuel Prices

The ascendance of fossil fuel prices is not the only ill effect of the drawdown of global fossil fuels. Fossil fuel energy markets are beginning to exhibit price trends that resemble some of the most ragged mountain ranges on the planet. Figure 1.3 displayed earlier exemplifies market volatility, as does figure 1.4, which charts the monthly price fluctuations of U.S. residential natural gas between 1981 and 2017.[28] There will always be peaks and troughs with residential gas prices because of the higher demand for heating fuel in the winter; however, peaks and troughs caused by supply-demand differentials in the winter and summer months are clearly becoming more pronounced. This trend suggests that the ability of natural gas suppliers to adjust supply to counteract intense demand fluctuations is increasingly limited by access to new natural gas finds.

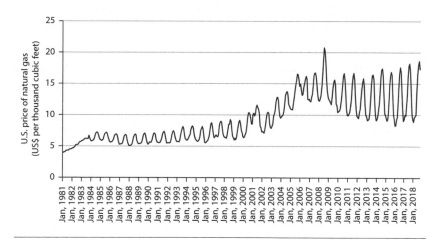

FIGURE 1.4 Capricious fluctuation of U.S. residential natural gas prices, 1981–2018; all prices in nominal U.S. dollars (U.S. Energy Information Administration, *Natural Gas*, accessed December 14, 2018, https://www.eia.gov/dnav/ng/hist /n3010us3m.htm.)

Such wild fluctuations of fossil fuel prices complicate management within any energy-dependent organization.[29] For power utilities, such fluctuations exacerbate purchasing risk and shift strategic attention from developing products (adding value) to managing costs (preserving value). For example, the public energy utility in Taiwan[30]—Taipower—and China's electric utilities[31] have both encountered the threat of operational insolvency because of coal and natural gas price fluctuations that are not compensated for in a timely fashion through permission to raise electricity rates. For utilities, some of this risk can be mitigated by investing in commodity futures to hedge against price increases, but all these financial instruments come at an added cost.

For energy-intensive firms, commodity price uncertainty wears a heavy cloak of risk. Energy price volatility can artificially mask competitive strengths and weaknesses and wreak chaos on profit and loss statements. Consider how price volatility in petrol impacts logistics companies such as DHL and Federal Express. Moreover, the added risk of doing business in a period of energy cost volatility forces firms to create rainy day funds—bolstering financial reserves and dampening the investments needed to expand capacity.

To mitigate the risks posed by unpredictable fossil fuel prices, most energy retailers and major consumers of energy are beginning to explore ways to diversify energy portfolios. Renewable energy technologies with their low variable cost profiles represent attractive portfolio additions. For fossil fuel energy systems, the price of fossil fuels represents a large portion of the cost of producing energy. Conversely, most forms of renewable energy exhibit cost profiles that are heavier on fixed investment but lighter on operating costs. For example, once a wind turbine is built, the cost of capturing wind and turning it into energy is very low and far more predictable.[32]

Trend 3. The Strategic Need to Diversify

The necessity to diversify the national fuel mix for the sake of national security has been reaffirmed throughout the history of humankind. The inhabitants of Easter Island purportedly caused their own demise by plundering the island's forests, which were their principal energy resource.[33] Germany's technological and military advantages during both world wars were undercut by shortages of fuels stocks,[34] and the Japanese suffered a similar fate in the Pacific war.[35]

However, the *economic* rationale for fuel mix diversification extends far beyond national security concerns. In modern times, when policymakers talk about the need to diversify a fuel mix, they are not thinking solely in terms of military necessity; they are also thinking in economic terms and point to the turmoil caused by the oil crises of the 1970s as events to avoid in the future. Indeed, even though the global oil supply network has diversified and the power of the OPEC oil cartel has diminished significantly,[36] there is still concern over the market manipulation of supplier nations.

The ambition to diversify is not just an aspiration centering on the unpredictability of oil cartels. Consider gas. In 2014, Russia's state-owned gas monopoly, Gazprom, threatened to significantly increase the price at which it sells natural gas to the Ukraine, catalyzing a spike in gas prices across Europe. This was not the first time this has happened. In both 2006 and 2009, Russia actually suspended exports of natural gas to Ukraine— during the winter periods—as a form of political pressure.[37] Russia's willingness to use its natural gas wealth for political purposes has caused a great deal of uncertainty throughout Europe, which depends on Russian natural gas for about a third of its gas imports.[38]

Technologically, there is also ample justification to diversify energy resources. Japan's predicament is one illustration. By having such a high reliance on nuclear power (30 percent) at the time of the Fukushima disaster, the nation exposed itself to amplified costs when it had to transition to other forms of energy as the nation's nuclear power plants were shuttered for inspection.[39] Simply put, systems that are severely impacted by trials and tribulations facing one energy technology exhibit risk that can be reduced through diversification of technologies.

Indeed, the admonition that one should refrain from putting all of one's eggs into one basket is very relevant in energy planning circles. In addition to the dangers of technological overdependence, there are risks associated with geographic overconcentration. With approximately four thousand rigs in the Gulf of Mexico, damage caused to U.S. oil infrastructure and the subsequent jump in oil prices in the United States as a result of hurricanes that pummel the eastern U.S. coast epitomize the risks associated with geographically concentrated energy infrastructure. In 2005, Hurricanes Katrina and Rita caused US$10 billion worth of damage to energy infrastructure. In the latter part of August 2017, Hurricane Harvey "shut down 22 percent of the nation's refining capacity, vitally disrupted the oil and gas transportation networks that deliver energy to much of the US, and caused damage to facilities that leaked more than a million pounds of dangerous air pollutants into communities around Texas."[40]

Another facet of the technological rationale for diversifying stems from the decentralized nature of many renewable energy technologies. These technologies are often operationalized through smaller power plants that can be dispersed around a nation. There are three important implications of this decentralization. First, these systems exhibit more resilience to national grid failures. Not all nations can boast resilient electricity grids—indeed, most do not. When power systems are centralized, a disruptive event (for example, a transformer malfunction or an extreme weather event) can knock out power across wide areas. To illustrate, in the U.S. state of Michigan, a late winter storm caused a blackout that affected over a million people, some of whom did not have power restored for four days.[41] In Costa Rica, over 5 million citizens were left without power after a transmission line in neighboring Panama collapsed.[42] With decentralized systems, blackouts do not have such widespread consequences. Second, in the event that smaller, decentralized systems fail, the impact on the grid will be far less damaging. When a nuclear power plant goes down, it can remove 1 gigawatt or more of generation potential from the grid. When a 3-megawatt wind turbine breaks down, the impact is negligible. Third, smaller decentralized systems engender a broader base of ownership, which helps attenuate the economic and social damage caused when mammoth-sized private utilities such as Enron in the United States and TEPCO in Japan take out huge swaths of innocents when they topple over.

Trend 4. Political Instability and Conflict due to Fossil Fuels

Returning to the topic of national security, history documents the precarious nature of overdependence on fossil fuel resources that are concentrated in a

TABLE 1.5 World's top 10 oil exporting nations, 2016

Nation	2016 crude exports (billion US$)	% of world total
Saudi Arabia	136.2	20.1%
Russia	73.7	10.9%
Iraq	46.3	6.8%
Canada	39.5	5.8%
United Arab Emirates	38.9	5.7%
Kuwait	30.7	4.5%
Iran	29.1	4.3%
Nigeria	27.0	4.0%
Angola	25.2	3.7%
Norway	22.6	3.3%

Source: Central Intelligence Agency, The World Factbook, 2017.

few nations. However, it is not just the economic ramifications of concentration of supplies that is of concern; there are also the geopolitical concerns.

Many nations that possess substantial reserves in oil, and to a lesser extent in natural gas, can be characterized as authoritarian-ruled economies. Consider the list of the world's top ten net oil-exporting nations in 2016 as depicted in table 1.5. In the *Economist*'s Democracy Index of 2016, six of the ten (Saudi Arabia, Russia, United Arab Emirates, Kuwait, Iran, and Angola) were classified as being ruled by authoritarian regimes, and two others—Nigeria and Iraq—were classified as "hybrid" regimes. Thomas Friedman refers to the link between oil profits and authoritarian regimes as the First Law of Petropolitics—as the price of oil goes up, the pace of freedom goes down.[43]

One characteristic of particular concern to nations that are importing oil and natural gas supplies from some of these authoritarian-ruled nations is the link between fossil fuel profits and the financing of terrorist activities. Thomas Friedman contends that Saudi Arabian oil profits fund up to 90 percent of the expenses incurred by radical Islamic organizations. Regardless of the percentage, it is hard to deny a link between oil profits and radical Islamic groups; after all, Osama bin Laden's wealth came from his family's construction company, which thrived on government contracts for oil infrastructure.[44]

Many energy security analysts in the United States and other allied nations have come to the realization that the high cost of defending access to oil and the links between oil profits and terrorism should catalyze a rethinking of energy security priorities. Former CIA director Jim Woolsey perhaps summarized this sentiment best: "we are funding the rope for the hanging of ourselves."[45] Columbia University's Jeffrey Sachs put U.S. efforts to preserve priority access to Middle Eastern oil into economic perspective: in 2007, "the United States spent an estimated US$572 billion on the military . . . and US$14 billion on development and humanitarian aid."[46]

Although the list of top natural gas–exporting nations boasts more politically stable economies, there is still strong representation from authoritarian-ruled nations. As table 1.6 illustrates, with the exception of European Union nations, Canada, Australia, and the United States, the list of top natural gas–exporting nations includes some politically unstable nations, some of which

TABLE 1.6 Top natural gas–exporting nations, 2017

Nation	Estimated exports (billion US$)	% of world total
Qatar	29.1	12.5
Norway	26.5	11.4
United States	22.3	9.6
Australia	20.5	8.8
Algeria	14.1	6.1
Canada	10.3	4.4
Malaysia	10.0	4.3
Indonesia	8.9	3.8
United Arab Emirates	8.4	3.6
Germany	7.3	3.1
Turkmenistan	6.6	2.8
Belgium	6.5	2.8
Nigeria	5.7	2.5
Saudi Arabia	4.9	2.1
Russia	4.7	2.1

Source: Data from Daniel Workman, "Petroleum Gas Exports by Country," *World's Top Exports*, June 11, 2018, http://www.worldstopexports.com/petroleum -gas-exports-country/.

have also been known to provide safe haven to terrorist organizations. The Council on Foreign Relations (CFR) describes Indonesia in this manner: "Indonesia, the world's most populous Muslim county, is a vast archipelago with porous maritime borders, a weak central government, separatist movements, corrupt officials, a floundering economy, and a loosely regulated financial system—all characteristics which make it fertile ground for terrorist groups."[47] Given this background, one can understand how many would perceive a shift from oil to natural gas to be a small upgrade in regard to enhancing geopolitical stability and why many energy analysts would prefer to see more technological diversity built into the supply process.

Trend 5. Improved Understanding of Fossil Fuel Health and Environmental Costs

Putting a price on the human health and ecosystem impacts associated with fossil fuel combustion is a notion that is fraught with contention; but there appears to be growing consensus among epidemiologists and ecological economists that the cost is too high to simply ignore.[48] Increasingly, academic analyses of energy costs attempt to improve the comparability of different policy options by seeking to "internalize" all relevant costs into the price of energy generated by different fuels and technologies.[49]

Internalization of all costs associated with conventional energy use poses some moral challenges. In China and India, the two greatest causes of premature death are cardiovascular and pulmonary diseases associated with coal-fired power plant emissions. According to a study undertaken by researchers at the University of British Columbia, in 2015, 2.2 million deaths were attributed to air pollution in these two nations.[50] There is a tendency to see these problems as issues that are found only in developing nations, but this is far from the truth. According to an MIT study, annually in the United States, pollution from vehicle emissions accounts for 58,000 premature deaths, and emissions from power plants are responsible for 54,000 premature deaths.[51] Increasingly, policymakers are beginning to realize the need to incorporate these hidden costs associated with fossil fuel combustion and use into the cost of energy services to better compare resource and technology options.[52] However, to accomplish this, we need to know how energy resources are converted to energy services (as in the "Sankey diagrams" produced by Lawrence Livermore National Laboratory for the United States),[53] and we need to evaluate technically and ethically complicated impacts of these conversions, such as the value of premature deaths. Indeed, should such outcomes even be valued economically? Then there is the issue of calculating the imminent and impending costs arising from future climate change.

Although there is still a great amount of fist shaking over what financial amounts should be assigned to the negative externalities caused by fossil fuel combustion and use, there is grudging consensus that some recognition of these costs is needed when planning energy futures. This debate has drawn mass media attention to the need to internalize costs that were, until recently, not part of the policy conversation. One such cost category involves the perils associated with amplified impacts of climate change.

Trend 6. Sobering Evidence of Climate Change Impacts

Currently the visual invasiveness of air pollution and smog arising from fossil fuel consumption draws the brunt of media attention. Images of cities such as Beijing and New Delhi cloaked in smog make far better press than do scientific updates on the progressive nature of climate change. Up until this point, for much of the general public, climate change has been a distant threat—a problem for future generations. However, as greenhouse gas concentrations continue to rise, we are increasingly reminded, in dramatic fashion, of the carnage to be wrought by climate change. This was publicly driven home through the images of Hurricane Sandy, which caused US$65 billion in damage in 2012 in the United States. In 2013, around the world, over 41 extreme weather events caused more than US$1 billion in damage each—a new annual record.[54]

In the United States, the National Oceanic and Atmospheric Administration (NOAA) has tracked the cost of extreme weather events for the past three decades. Overall, the United States experienced 218 extreme weather events between 1980 and 2017, incurring total costs of US$1.3 trillion.

In 2017 alone, 16 of these weather-climate disasters exceeded $1 billion, totaling a new record for damages—US$306 billion.[55] Obviously not all are the result of climate change, as storms were occurring long before humankind began spewing out greenhouse gas emissions. However, it would be reasonable to assume that climate change plays a role if these extreme weather events are intensifying. The NOAA data can help us assess this question because they also track extreme weather events that incur over US$1 billion (inflation-adjusted) in damages. Between 1980 and 2016, the annual average has been 5.5 events (Consumer Price Index (CPI)–adjusted). However, the annual average for 2012–2016 was 10.6 events (CPI-adjusted). One contextual nuance that is often lost in numerical analysis is that each year capital investment in coastal areas increases, so each year there are more assets at risk as climate change progresses.

The assessment reports issued by the Intergovernmental Panel on Climate Change (IPCC) in 1990, 1996, 2001, 2007, and 2014 progressively reflect how climate change modeling and data are increasingly better able to characterize developments under various greenhouse gas concentration scenarios. Yet one suspects that it is not greater scientific certainty regarding the threat of climate change that is primarily responsible for engendering growing support for renewable energy; rather, it appears to be the real and present cost of these fossil fuel harbingers of climate change that catalyzes change. Policymakers are finally coming to the realization that the distant costs attributed to climate change, which were projected ten years ago, are now imminent. Tomorrow is today, and the day after tomorrow could be worse. When you give a typical politician a choice between doing nothing or incurring a cost today to offset a greater cost in the distant future, most politicians will normally embrace the "do nothing" option as if it were a puppy at a fundraising gala. However, when fiscal budgets become impaired by rising climate change adaptation costs, the "do nothing" option quickly loses its appeal.

Trend 7. The Contested Politics of Nuclear Power

In addition to fueling support for renewable energy, the first six trends also form a foundation for support for a potential expansion in nuclear power capacity. As testament to this view, in 2008 the World Nuclear Association projected that installed nuclear power capacity would grow between 38 percent and 208 percent by 2030.[56] In 2009, the International Atomic Energy Agency reported that more than sixty nations were investigating the adoption of nuclear power.[57]

Then on March 11, 2011, the nuclear renaissance came to a screeching halt following the disaster at the Fukushima Daiichi nuclear power plant, caused by a tsunami that hit Japan following the Great East Kanto Earthquake. Many nations immediately put on hold their nuclear power adoption and expansion plans and launched reviews of their nuclear power programs. Germany decided to hasten plans to phase out nuclear power as a result of its review. Nuclear power expansion initiatives in the United States, Canada, Taiwan, the United Kingdom, and even France have come under enhanced scrutiny. In Japan, debate over whether to reinitiate the nation's nuclear power program has been protracted,[58] with recent inroads made by renewable energy technologies complicating the picture.[59] In January 2018, an influential group led by former prime ministers Junichiro Koizumi and Morihiro Hosokawa unveiled a proposed bill to begin an immediate total phaseout of the nation's nuclear reactors.[60]

This is not to say that all nations have given up on nuclear power. China, South Korea, and India are three prominent nuclear nations that are resolutely proceeding with nuclear power expansion plans despite the political fallout from the Fukushima disaster.[61] Other nations, such as Japan, France, the United Kingdom, Canada, Russia, and the United States, seem to be mired in a game of "wait and see." As of January 2017, 450 nuclear power reactors were operating in thirty countries and 60 more reactors were under construction in fifteen countries.[62] Asia is a hotbed for development, with 40 reactors under construction (including 20 in China alone) and plans in the region for 90 more.[63] To the glee of many in the nuclear power industry, some prominent environmental scientists, such as Columbia University's James Hansen, lend credence to expansion aspirations in advocating that nuclear power is a necessary evil to abate climate change.[64]

Developmental hot spots aside, the accident in Fukushima has raised the perceived risk profile of nuclear power in many nations and has prompted some nations to consider renewable energy as a lower-risk strategy for mitigating climate change. Other substantial problems surrounding nuclear power adoption include the proliferation of nuclear weapons, financial viability, waste storage, and operational performance. For many of these reasons, the industry often faces staunch resistance from the general public when new reactors and nuclear-related facilities are proposed.[65]

Meanwhile, there is an additional concern associated with nuclear power that is not often spoken about—resource scarcity. Industry supporters argue that nuclear power plants use very little uranium to produce energy, and therefore both the demand on this finite natural resource and the waste produced are comparatively insignificant.[66] However, if one were to apply to uranium the same reserve-to-production ratio analysis that was conducted in table 1.3 for fossil fuels, the threat of scarcity should be apparent (see table 1.7).

Some have argued that uranium scarcity should not be considered a constraint for nuclear power development because there are other sources of nuclear power, such as thorium, which if adopted for the nuclear fuel cycle would provide a much more abundant source of energy.[67] This, however, brings to light the more compelling reason why nuclear power development has been stagnating—cost. Simply put, when the costs of plant construction, plant operations, waste management, and plant decommissioning are built into economic assessments, the case for constructing new nuclear units often cannot be defended.[68] While arguing in favor of completing two half-built nuclear units (the only two units under construction

TABLE 1.7 Reserve-to-production estimates for uranium

	Proven reserves	Current production	Reserve-to-production ratio (years)		
			0% annual production growth rate	1.6% annual production growth rate	2.5% annual production growth rate
Uranium	4,743,000 tons (at US$130/kgU)	40,260 tons	118	67	56

Source: B. Sovacool and S. Valentine, *Sounding the Alarm: Global Energy Security in the 21st Century*Energy Security (London: Sage Library of International Security, 2013), 3.

in the United States today), one of this book's authors acknowledged that a case for completion from scratch could not be justified today on traditional economic grounds.[69] Given the commercial progress of renewable technologies and the availability of cheap natural gas, nuclear energy simply does not make economic sense with its current technological status. However, a phaseout will not be without costs. As the nuclear power industry declines, it discourages the maintenance of an important nuclear weapon antiproliferation safeguard: "a bunch of smart nuclear scientists and engineers" who can conduct the inspections needed to keep the world safe from nuclear weapons.[70]

Trend 8. Innovations in Performance and Cost Within the Renewable Energy Sector

The economic appeal of coal has lost its sheen. Without a doubt, power from a coal-fired power plant that uses brown coal (lignite) and operates without any emission control equipment is cheap—certainly, less expensive than power generated by wind, solar, wave, and tidal power, in most cases. However, in all major industrialized nations—and increasingly in developing nations—coal-fired power plants operate under stringent emission control standards. Chinese coal-fired power plants are an example. In a comparative study, Greenpeace reports that in many cases Chinese coal plant emission standards are actually higher than Japanese emission standards.[71] Most new coal-fired power plants must incorporate costly technologies for reducing the amount of pollution emitted. Moreover, most major nations have turned away from lignite coal, which causes the most damage from an environmental perspective. The global standard is increasingly the harder black coal (bituminous), which is more expensive on a per-kilowatt-hour basis. The consequence of these developments is that advanced coal-fired power plants are producing energy at increasingly high costs, and this is narrowing the economic divide between coal-fired power and some of the renewable energy alternatives.

Just how significant is this cost convergence? Figure 1.5 portrays the levelized cost of electricity for new renewable power plants in 2010 and 2017, compared to the range for fossil fuel cost. As the data suggest, coal's reign as

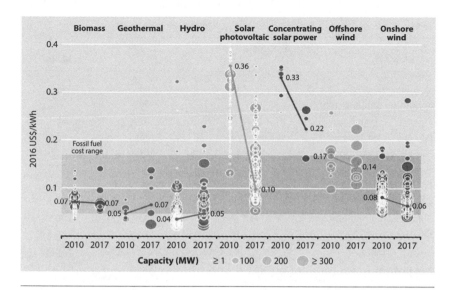

FIGURE 1.5 Global levelized cost of electricity for new utility-scale power plants, 2010–2017 (International Renewable Energy Agency [IRENA], 2018.)

the cheapest form of energy provision is under siege. Electricity generated from biomass, geothermal energy, hydro, solar photovoltaic (PV), and wind power is competitive with electricity generated from fossil fuels. And if the cost of concentrating solar power continues its rapid decline, it, too, will be competing in the fossil fuel sandbox. In short, energy markets are in the process of being disrupted by renewable technologies.

Table 1.8 presents a similar levelized cost analysis (not including internalized environmental and social costs) for some of the major power generation technologies.[72] Solar PV in particular has surprised most analysts over the past five years. According to a recent study by researchers at the National Renewable Energy Laboratory, between the first quarter of 2016 and the first quarter of 2017, the cost of utility-scale, fixed tilt solar PV systems declined by a whopping 27 percent.[73] Between 2016 and 2018, the cost of utility-scale, fixed tilt solar PV systems declined 42 percent.[74] In short, given the data in table 1.8, if you were an energy planner for a given nation, which technology would you choose?

Indeed, looking to the future, the most recent projections from IRENA depicted in figure 1.6—reflecting much of the data and potential

TABLE 1.8 Levelized costs of major energy technologies, 2016

(US$/MWh)	Low	High	Mean
Wind power	32	62	47
Solar PV—thin film utility	46	56	51
Solar PV—crystalline utility	49	61	55
Gas combined cycle	48	78	63
Natural gas reciprocating engine	68	101	84.5
Biomass direct	77	110	93.5
Geothermal	79	117	98
Coal	60	143	101.5
Solar PV—community	78	135	106.5
Nuclear	97	136	116.5
Solar PV rooftop	138	222	180
Gas peaking	165	217	191

Source: Lazard's Levelized Cost of Energy Analysis—Version 10.0, accessed November 5, 2017, https://www.lazard.com/media/438038/levelized-cost-of-energy-v100.pdf.

Note: Excludes subsidies.

*MWh = megawatt hour.

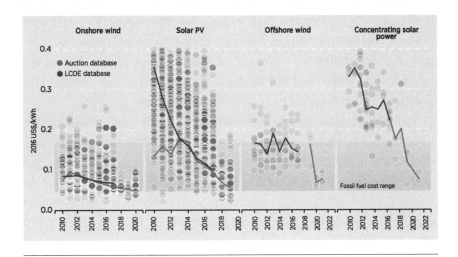

FIGURE 1.6 Global levelized cost of electricity for various sources of renewable energy, 2010–2022 (International Renewable Energy Agency [IRENA], renewable cost database and auctions database, December 2017.)

for learning we've presented above—suggest that the cost of renewable energy will be reduced by *half* again by 2020. As Professor R. Andreas Kraemer observed when reviewing these figures, "This implies a rapid death for fossil and nuclear energy, and likely spillover of the disruption from the power sector to transport (driving out diesel and petrol), cooking (driving out kerosene in developing countries), power generation (driving out diesel from power generation), and heating in homes, business and industry."[75]

Trend 9. The Rise of Government and Market Support for Renewable Energy

Unfortunately, continued political support for conventional energy has never been just about finding the least-cost solutions. As fossil fuels ascended to dominance within energy markets, intricate supply chain webs developed in support of these energy services, fueling the rise of a network of financially powerful special-interest groups. This development has culminated in the creation of a widespread network of fossil fuel advocates who are wealthy, politically connected, and vehemently opposed to change. Facilitating a transition away from conventional energy is not just about having cheaper technology; it is also about overcoming resistance from these special-interest groups.

Energy policymakers in many nations are increasingly aware of the uphill battle that renewable energy firms have been fighting. In many nations, the "goods" of renewable energy and the "bads" of conventional energy technologies have combined to prompt policy support for renewable energy. The result has been an increase in renewable energy incentives (see figure 1.7) that center on subsidization via feed-in tariffs, mandatory purchase programs, production tax incentives, and development grants.

As an indication of the changing policy landscape, as recently as 2008, the bulk of energy subsidies was being directed to the fossil fuel industry. In the United States, this changed in 2011, when government agencies spent two-thirds (US$16 billion) of their total energy subsidies (US$24 billion) on renewable energy and energy efficiency initiatives.[76] As figure 1.7

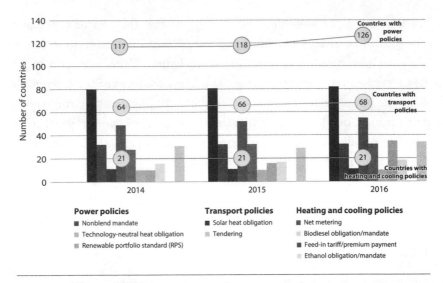

FIGURE 1.7 Growth of renewable energy policies, 2014–2016 (Redrawn from Renewable Energy Policy Network for the 21st Century [REN21], *Renewables 2017 Global Status Report*, http://www.ren21.net/gsr-2017/.)

suggests, worldwide, the policies that are being put in place are exhibiting staying power.[77]

As government support for renewables congeals worldwide, investors and private industry are getting the message. Record sums of investment have been pouring into the renewable energy sector. Although investment tailed off somewhat in 2016 as a result of an economic slowdown in many emerging economies, new investment in the renewable energy sector advanced again in 2017. Developing countries, in particular China, have extended their lead over developed countries (see figure 1.8).[78] And for the eighth consecutive year, investment in renewable energy surpassed investment in conventional energy technologies. According to the Renewable Energy Policy Network (REN21), in 2017, renewable energy investments (excluding hydro) constituted 58.4 percent of total global investment in new capacity.[79] At long last, renewable energy firms are experiencing a continuous influx of support, and this translates into the declining cost profiles outlined earlier.

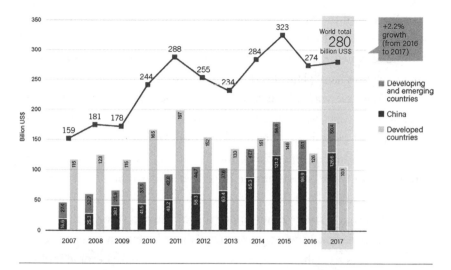

FIGURE 1.8 Investment in renewable energy, 2007–2017 (Renewable Energy Policy Network for the 21st Century [REN21], *Renewables 2018 Global Status Report*, http://www.ren21.net/gsr-2018/.)

Trend 10. First-Mover Advantages amid a New Energy Boom

As a global energy transition gains steam, more industry analysts are beginning to recognize that the actions required to facilitate such a transition—revamping electricity grids, restructuring energy networks, reengineering energy systems—all represent stellar economic opportunities. A report commissioned by the German government concluded that the size of the global market for environmentally friendly power generation and storage was approximately €313 billion in 2011 and would exceed €1 trillion by 2025.[80] As former U.S. president Obama emphasized in a speech given in Iowa in 2009, "The nation that leads the world in creating new sources of clean energy will be the nation that leads the twenty-first-century global economy."[81] In 2015, he spoke again about the changing competitive tides: "For decades, we've been told that it doesn't make economic sense to switch to renewable energy. Today, that's no longer true. . . . Solar isn't just for the green crowd anymore—it's for the green eyeshade crowd, too."[82]

Energy is an industrial sector that is on the verge of a complete makeover—a makeover that will create new global corporate powerhouses.

Matt Richtell of the *New York Times* describes the scene in this way: "Out of the ashes of the Internet bust, many technology veterans have regrouped and found a new mission in alternative energy: developing wind power, solar panels, ethanol plants and hydrogen-powered cars."[83]

As is the case with all business sectors,[84] there are first-mover advantages to firms that can secure early market share leads. Pioneering companies benefit from competitively mature supply networks and economies of scale in production, operations, and marketing.[85] First movers also have more time to work out systematic bugs that undermine profitability and reputational value.

It is clear that first-mover advantages are already accruing to firms in the clean energy sector. In wind power, Danish and German firms were first movers; and as a result, Vestas, Bonus, and Siemens have become industry leaders. In solar power, Chinese firms have come to dominate the market, thanks in part to staunch government support but mostly because many of the Chinese solar PV firms could draw on the expertise of Chinese engineers who cut their teeth by working with solar PV technology in the consumer electronics sector. In wave power technology, true dominance has yet to be established, but one can surmise that firms in Portugal, the United Kingdom, Japan, Australia, and the United States stand to gain from early domestic support. Nevertheless, the emergence of renewable energy powerhouses represents just the tip of the iceberg in regard to the global energy transition. Smart meters, energy storage, grid management technologies, smart electric charging infrastructure, and energy-efficient technologies are all sectors that will become major economic players as the transition progresses. It has been estimated that the entire clean-tech sector could generate upwards of €4 trillion in sales by 2025.[86] There is a piece of the pie for everyone at this early stage.

LIMITED EVIDENCE OF IMPACT OF THE TEN TRENDS

These ten dominant trends should be fueling a boom period for renewable energy development, but the statistics paint a different picture. By December 2016, the contributions from renewable energy had risen to only

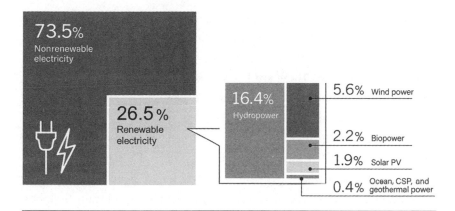

FIGURE 1.9 Renewable energy share of global electricity production, 2017; CSP = concentrated solar power (Renewable Energy Policy Network for the 21st Century [REN21], *Renewables 2018 Global Status Report*, http://www.ren21.net/gsr-2018/.)

26.5 percent of the global energy mix, a plodding increase from 19 percent in 2011. Moreover, over 62 percent of the renewable power generated came from hydropower. As figure 1.9 illustrates, the "modern renewables" that are supposed to drive the transition—wind power, solar PV, biomass, geothermal, and others—contributed only 10.1 percent of total electricity consumption.[87]

This begs the question, given the strength of the forces for change (as suggested by the ten trends), what is deterring the diffusion of renewable energy? In chapter 2, six factors are introduced that play a major role in inhibiting the scale and pace of change. If an energy transition is to be expedited to avert the worst impacts attributed to climate change and afford a nation a degree of first-mover advantage, policymakers in a given nation will need to develop strategic policy responses to reduce these impeding forces. To reiterate an earlier point, energy politics is not just about energy economics.

This "rule-of-thumb" sets the stage for ensuing chapters. Chapter 3 examines how climate change science has impacted energy market dynamics and offers approaches for synthesizing competing perspectives on climate change. Chapter 4 investigates strategies that governments could use to manage technological evolution and the profound uncertainties that surround energy transitions. Smart grids, homes, businesses, and transport are being empowered by electrification, digitalization, and decentralization, and they are increasingly interconnected through novel forms of collaborative consumption. These are key ingredients of a recipe for replacing our old inefficient diet of fossil fuels with a new, responsive, and inclusive smorgasbord of clean-tech selections.

Chapter 5 investigates ways to foster and finance an extreme energy infrastructure makeover. Beginning with an overview of international investment commitments that highlights the need for cooperation, the chapter then evaluates projected investment costs to support a full transition, how infrastructure investments can be optimized, and the role that policy can play in expediting market activity. Chapter 6 extends the examination of policy drivers by documenting the full array of policy instruments available to policymakers. After describing the theory and logic of policy intervention, it focuses on documenting, categorizing, and evaluating the utility of a vast array of policy instruments that are used around the world for promoting market and behavioral change.

Chapter 7 reflects on the challenge of altering consumer behavior to drive demand from the bottom up. The reader will see the array of factors that influence human behavior and stymie change. Accordingly, the crux of this chapter centers on strategies to help foster behavioral change. Chapter 8 focuses on altering approaches to governance. It begins by defining governance (and distinguishing it from government or policy) and explaining why energy and climate change mitigation present daunting governance problems. It focuses on six elements that influence energy planning—policy trade-offs, transboundary externalities, the political economy of trade and intellectual property, subsidies, the challenge of wicked problems, and the rapidity of temporal and technological change. The chapter ends by discussing how polycentric governance approaches may resolve paralysis at the national level. Finally, chapter 9 concludes by integrating and aligning recommendations from all previous chapters to ensure that national policy strategy is firing on all cylinders—optimizing effectiveness and delivering maximum value to the nation.

This book is primarily about what can be achieved by policymakers who wish to lead change. Although it must be acknowledged that politics and policy can rarely be separated when it comes to an area such as energy planning—especially where trillions of dollars are at stake—the philosophical premise of this book is that change is possible, even in the face of political opposition, with a cohesive, polycentric strategy. We have seen examples in the past where technological incumbents were unseated from market leadership by technology that better served emergent social needs. As a case in point, consider how often you use your telephone at home now. Even in the energy world, we have experienced large-scale shifts from wood to coal, coal to oil, and oil to natural gas. We have seen ground transportation systems evolve from horses to horse-drawn carriages, eventually ceding way to horseless carriages and then automobiles, trolleys, trains, and buses.

Now we face yet another transition—one that will alter fortunes and radically enhance efforts to establish a more sustainable economic development trajectory. In this chapter, we reviewed ten forces portending change in the energy sector. In the next chapter, we will consider the barriers that need to be overcome to bring added certainty to this transition. That cup of liquid we introduced at the beginning of the chapter is indeed coming off the table, so let's try to manufacture a safe landing spot.

2

Sneak Preview of the Challenges to the Energy Transition

Imagine a densely populated nation—Seusstopia—that relies on one dominant technology for its well-being. At first, the technology helps foster huge development leaps forward, and throughout the nation citizens spend their days swaying together atop lush green hills holding hands and singing lyrical couplets. As time goes on, use of this technology gives rise to widespread ecological degradation that has not been seen since the days of the Lorax. A few citizens find themselves overcome by a pall of depression. They no longer feel compelled to scale the heights to join their fellow citizens in song. Instead they stand silently at the bottom of the hills, passively viewing the celebrations of their fellow citizens in thought-induced silence. A few long voices articulate the thoughts of many—*Is this progress?* Yet the population grows, the technology is increasingly deployed, and the degradation continues. First the songs on the hill lose their impassioned resonance, then clashes between citizens occur, and then disenfranchised individuals begin to commit acts of sabotage on persons and property. People fall ill from this toxic environment, and the technology becomes the leading cause of human despair.

Imagine that amid this scenario, a lone cart trundles down the dusty and now treeless road that winds into town. The lady in the cart announces to the citizens that beneath the cloak that covers her cargo is a mix of wondrous new technologies that will attenuate the degradation to the environment, restore economic opportunity to those in need of work, and provide the same services as the old technology at a lower cost. Be honest: would you not consider any citizen who refused to embrace this new arrival to be irrational?

This is where we now stand with energy sector governance. As we saw in chapter 1, there are at least ten significant trends that, in aggregate, portend conditions that compel a shift away from carbon-based energy forms to renewable energy technologies. In undertaking such a transition, we will be averting ecological, social, human health, geopolitical, and economic damages associated with our continued reliance on fossil fuel energy. However, instead of leading to a global stampede toward renewable energy, statistics show that this much-anticipated transition is materializing in sloth-like fashion. Why?

In this chapter, we will introduce six prominent and admittedly interconnected challenges to a transition away from fossil fuel energy technologies: (1) scientific uncertainty regarding climate change impacts, (2) resistance from powerful entrenched interests, (3) difficulties in fostering and financing a transition, (4) difficulties in reshaping market dynamics with the right policy mix, (5) consumer apathy, and (6) failures in politics and governance. These challenges are formidable hurdles to elevating the scale and pace of energy transition and will be examined in greater detail in the chapters that follow. By reflecting on the causal forces that give rise to these challenges, we will be in a position to consider strategic approaches to overcoming them and, in the process, identify strategies to fast-track a switch to cleaner forms of energy. Underpinning our analysis is this truth—the obdurate beast that we face is us.

SCIENTIFIC UNCERTAINTY REGARDING CLIMATE CHANGE IMPACTS

In the absence of a climate change threat, a transition away from low-carbon energy systems would likely be a gradual affair, prolonged for decades if not centuries by aggressive fossil fuel discovery initiatives—initiatives that would extend proven reserves and bolster the capacity to supply energy needs through fossil fuels. As testament, it took coal over 170 years to reach the stage where it was delivering 25 percent of our global energy needs.[1] However, climate change significantly alters the desirability of a slow evolution. No scientist of any repute would deny that preserving the status quo—continuing to generate energy predominantly through fossil fuel energy systems—is a recipe for ecological, social, and economic woe.[2]

Most climate scientists would agree with the Fifth Assessment Report (AR5) of the International Panel on Climate Change (IPCC), which concludes that keeping atmospheric emissions within an upper limit of 450 parts per million (ppm) of CO_2-equivalent greenhouse gases by 2100 is *likely* needed for humanity to have any chance of staving off a global temperature rise that exceeds 2°C.[3] Yet, as AR5 points out, "Baseline scenarios (scenarios without explicit additional efforts to constrain emissions) exceed 450 parts per million (ppm) CO_2eq by 2030 and reach CO_2eq concentration levels between 750 and more than 1300 ppm CO_2eq by 2100."[4] In other words, we have some work to do. Ominously, even if we hit these bold reduction targets, CO_2 might actually have to be extracted from the atmosphere to ensure that a temperature rise does not breach the 2°C target.[5]

To complicate matters, not everyone agrees that this 450 ppm target will rein in temperature rise to under 2°C. For example, prominent former NASA scientist James Hansen and a group of nine other academics published a paper in 2008 arguing that 350 ppm was a more appropriate upper limit to set in order to fend off ecological disaster.[6] This 350 ppm target has also been embraced by environmentalist Bill McKibben and colleagues, who have established a website to promote the imperative of staying within 350 ppm (www.350.org).

Moreover, there is a significant amount of dissent over the *pace* at which greenhouse gas emissions need to be reduced to stay within the 450 ppm limit. The reason is that greenhouse gas molecules in the atmosphere have molecular shelf-lives. A CO_2 molecule that makes it into the atmosphere today will not degrade tomorrow. Rather, it will remain in the atmosphere for decades (and for other greenhouse gas molecules, even centuries and millennia); and although the potential of a molecule to absorb heat degrades over time, these CO_2 molecules will continue to exhibit heat retention prop-erties.[7] Because the science regarding the degradation of heat absorption by atmospheric greenhouse gas molecules is imperfect, some scientists argue that it would be prudent to decrease greenhouse gas emissions as early as possible to sidestep the threat associated with all-out efforts to hit a target that turns out to be insufficient.

In summary, then, although the science of climate change is improving and our understanding regarding systemic feedbacks is becoming more pre-cise, there is still disagreement over what we need to accomplish. In 2006, Nicholas Stern and his team released a report arguing that the upper limit should be 550 ppm.[8] Two years later, Hansen and his colleagues released their treatise arguing that 350 ppm should be the upper limit. Since then, the undeniably pluralist IPCC, which operates based on universal consen-sus when it comes to scientific judgment, has settled on the 450 ppm tar-get.[9] With such difference of opinion regarding what actually needs to be achieved, it should come as no surprise that disagreement over the pace at which greenhouse gas reductions need to be achieved also exists. In short, uncertainty is rife when it comes to climate change science, and scientific uncertainty produces caution in policymakers when it comes to altering a status quo that includes some powerful stakeholders.

In chapter 3, we will examine climate change science in more detail. The chapter begins by addressing and refuting prominent points of contention raised by climate change skeptics. The reader will learn why sunspots and volcanic activity are minor players and why a vengeful Mother Nature is also not to blame. It will then explore the potential of extreme impacts attributed to accelerated concentrations of greenhouse gases. This overview of worse-case scenarios is not intended to create a sense of alarm but to underline an important fact: failing to address climate change has a far greater downside for disaster than any economic costs incurred to mitigate such possibili-ties. The middle part of chapter 3 considers the drivers underpinning public apathy toward this exigent threat and investigates how scientific uncertainty is enabling advocates of the status quo to cultivate a climate of confusion. This discussion sets the stage for the last part of the chapter, which evaluates international mitigation efforts and puts forth prescriptions that the average citizen can follow to inspire politicians and policymakers to act.

RESISTANCE FROM POWERFUL ENTRENCHED INTERESTS

In the waning days of the Pacific War, it was painfully obvious that the sun had set on Japan as the Pacific's military powerhouse. The nation still boasted a high number of willing conscripts; however, it had run out of the resources necessary to render these conscripts effective in battle. U.S. naval forces had been effective in blockading Japan's Southeast Asian oil supply routes, and as a result, Japan had virtually exhausted its cache of oil to power its ships, planes, and tanks.[10] Yet Japanese leaders refused to accept defeat. Instead, they fabricated a series of irrational strategies. When the nation ran out of fuel to power routine air missions, they embarked on a massive campaign to dig up the roots of pine trees to extract biofuel. Hundreds of thousands of roots were needed to support a single air mission. On another front, they enacted a strategy to persuade inexperienced young pilots to sacrifice their lives by turning airplanes into bombs—the birth of the *kamikaze*.[11] Similarly, lacking the fuel to allow the navy to defend sea routes, Japanese leaders coerced young naval officers to sacrifice their lives by ramming enemy ships with submersible craft that were equipped with warheads at the front—the birth of the *kaiten* submarine. These initiatives would not win Japan a war that it was sorely losing.

Decisions made during the collapse of Japan as Asia's military powerhouse epitomize the type of irrational thinking that once-powerful countries, organizations, and individuals exhibit when facing the prospects of demise. The decision-making process becomes irrational because decision makers who were once powerful have immense difficulty detaching themselves from the strategies that had at one time propelled their rise to prominence. This ideological "stickiness" presages a paradigm shift.[12]

In the modern energy sector, there is a significant amount of evidence that proponents of conventional energy are entertaining kamikaze strategies of their own. There's good reason for this. Energy is big business, and conventional energy firms are still highly profitable and awash with cash. One can imagine the ideological hurdles that executives from ExxonMobil face when profits continue to amass as a result of their unfaltering conventional energy exploitation strategy. In 2013, Exxon earned US$32.58 billion in profits. In 2014, the firm earned another US$32.52 billion in profits; and in

2015 and 2016, profits still amounted to US$16.2 billion and US$7.8 billion, respectively. All the while, the firm has been leveraging high oil prices to expand production capacity. In 2016 alone, Exxon added 250,000 oil-equivalent barrels per day of production capacity.[13] Conventional energy is perhaps the world's largest cash cow. In 2014, it has been estimated that revenues from the oil- and gas-drilling sector alone amounted to between 4.6 and 6.5 percent of global GDP.[14] Energy expert Travis Bradford estimated that the entire sector accounts for 7 to 10 percent of global GDP.[15] When compared to global investment in renewables, until recently fossil fuel investments continued to wash into the sector. For 2013, firms and financiers invested roughly three times *more* in the fossil fuel sector (mostly for oil and gas ventures) than in renewable energy technology, as figure 2.1 indicates.

Since then, tides have changed somewhat. Investment capital continues to flow into renewable energy, but investments in fossil fuel systems have tapered off, resulting in a new world order where renewable energy technologies are now attracting the bulk of investment (see figure 2.2).[16]

Nevertheless, because of the sheer scale of investment that fossil fuel technologies have enjoyed, there is now an enormous amount of sunken investment. In 2016, the International Energy Agency (IEA) estimated that US$1.7 trillion was invested in energy infrastructure, with the vast majority going to the extraction and transportation of fossil fuels, oil refining, construction of fossil fuel–fired power plants, and establishment of electricity infrastructure.[17] It is not inconceivable that total sunken investment in conventional energy infrastructure is in the ballpark of US$20 to US$30 trillion—equating to approximately one-quarter to one-third of global GDP in 2018.

With this much money to lose in sunken investment and so much to gain in terms of ongoing profits, one can safely surmise that powerful stakeholders from the conventional energy sector are not motivated to change course. As we emphasized in a previous book, these stakeholders are not just found in boardrooms:

> Entrenched interests are everywhere. Economic interests committed to energy production begin at the extraction phase—exploring and drilling for oil and natural gas, mining coal and uranium, cultivating biomass, building dams, and harvesting wind and solar energy for power

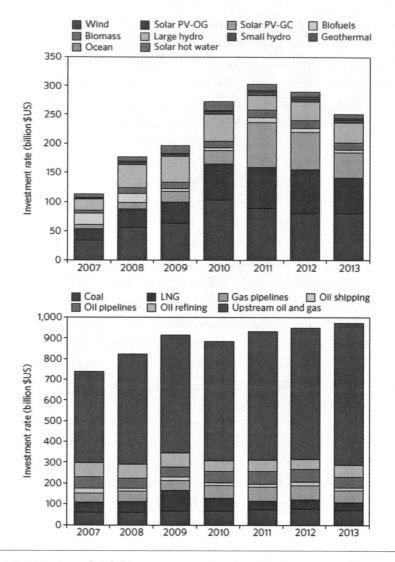

FIGURE 2.1 Annual global investment rates in selected energy systems, 2007–2013; OG = off grid, GC = ground coverage, LNG = liquid natural gas (P. C. Stern, B. K. Sovacool, and T. Dietz, "Towards a Science of Climate and Energy Choices," *Nature Climate Change* 6 [June 2016]: 547–555.)

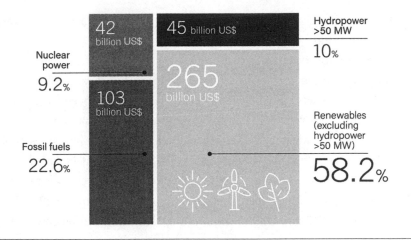

FIGURE 2.2 Global investment in new power capacity, by type, 2017 (Taken from Renewable Energy Policy Network for the 21st Century [REN21], *Renewables 2018 Global Status Report*, http://www.ren21.net/gsr-2018/.)

production. At the manufacturing stage—refining oil, processing natural gas, cleaning coal, pelletizing and refining biomass, and transporting energy commodities—a prodigious number of activities take place and each activity is supported by embedded investment, contractual commitments to workers and political ties. There are also thousands of energy-related companies at later stages of energy conversion and use. For instance, the United States has more electric utilities than fast food restaurants. Similar to the upstream situation, each of these downstream enterprises has entrenched investments that engender opposition to change. So eliciting change is not just about convincing behemoths such as the Exxon Mobils of the world to embrace wind power. The challenge is less akin to turning a supertanker and more akin to trying to align a cluster of marbles atop a table on a sailboat in rough seas.[18]

The pervasiveness of conventional energy within our global economic system has also given rise to social and technological locks that stymie change. For example, it has been estimated that 630 million workers across the globe are engaged in activities that are directly connected to energy extraction, production, or consumption.[19] Transitioning away from carbon-based energy systems threatens the livelihoods of these workers around the world.

Moreover, the companies that employ these workers have invested in the infrastructure necessary to deliver conventional energy as efficiently as possible. The networks that currently exist, and that have been largely supported through government subsidies, provide conventional energy firms with competitive advantages that new market entrants are hard-pressed to overcome. According to a comprehensive 2016 study that included costs arising from climate change impacts, air pollution, and vehicle externalities (such as congestion), fossil fuel subsidies amounted to US$4.9 trillion worldwide in 2013 and US$5.3 trillion in 2015. Even when narrowing the cost estimate to only supply costs and general consumer taxes, the study concluded that subsidies in 2015 were in excess of US$1 trillion.[20]

To preserve the golden calf, conventional energy firms and special-interest groups funded by energy firms have cobbled together aggressive market defense strategies, and it has been shown that many are not averse to sponsoring misinformation campaigns when necessary.[21] Indeed, misinformation—perpetuated by fossil fuel interests—that disparages and distorts the science underpinning climate change has prompted former NASA climatologist James Hansen to argue that CEOs of energy firms that are found guilty of disseminating such information should be tried for "*crimes against humanity*."[22] The resources behind these attempts to mislead the public are not trivial; Robert Brulle laboriously examined the tax records of "counter climate change movements" and noted more than 91 institutions funded by 140 different foundations from 2003 to 2010.[23] Together, these 91 institutions reported an annual income of more than US$900 million. Such efforts have been shown to artificially polarize climate change discussions and shape policy in the United States.[24]

Misinformation is not the only strategy employed by conventional energy interests to inject uncertainty into an otherwise certain shift away from carbon-based energy systems. Another strategy (or perhaps "Hail Mary" is the better term) is the advancement of solutions that would allow portions of the status quo to be preserved. One prominent solution of this ilk is carbon capture and sequestration (CCS), which is in the family of approaches called carbon dioxide removal (CDR).

Proponents contend that "as a society, we need to better understand the potential cost and performance of CDR strategies for the same reason that we need to better understand the cost and performance of emission mitigation strategies—they may be important parts of a portfolio of options to stabilize and reduce atmospheric concentrations of carbon dioxide."[255] Some further argue that CCS might be necessary as an interim solution because many nations will not have the financial capacity to shutter coal-fired power plants that are still being amortized. For example, the rapid buildout of coal-fired power plants that has occurred in China over the past two decades has left the nation with a fleet of relatively new, advanced coal-fired plants.

Yet, in many ways, CCS is a kamikaze strategy. It is a solution with dubious logistical plausibility that ultimately serves to prolong humanity's dependence on scarce, finite energy resources. It's been estimated that if all of the world's CO_2 emissions could be captured and transported in liquid form for geological sequestration, about 33 million oil barrels per day would have to be transported and sequestered. In research that centered on evaluating whether so much sequestered carbon could be stored worldwide, experts in Europe concluded that all large potential storage sites would reach capacity after thirty-six years.[26] In other words, there simply is not a big enough geological rug under which to sweep all the emissions associated with conventional energy combustion.

While CCS might prove a necessary stopgap on a limited scale to buy nations some time to finance the transition, it is hard to embrace CCS as a universal solution to greenhouse gas emission reductions given the scale of investment that would be required. Overall, in order to entrench such a technological solution, global investment requirements would run in the trillions of U.S. dollars, far more costly than pursuing a renewable energy path[27] (refer back to table 1.8). Whether the promise of CCS continues to dampen the transition toward renewable energy technologies depends on the rate at which fossil fuel prices continue to escalate, the pace at which the price of renewable energy continues to decline, and whether coal-fired power and CCS technology can evolve to a point of competitive viability with renewable energy and storage.

Casting a dark hue over the viability of CCS, renewable energy experts Elliston, Diesendorf, and MacGill argued as far back as 2013 that current renewable energy configurations are 100 percent feasible and can be less costly than many advanced fossil fuel configurations before options such as CCS are even factored into the equation.[28] A substantial body of published research confirms that at low penetration levels, storage of solar and wind power flows (<10 percent) carries minimal integration costs. For example, a study by researchers at three Department of Energy National Laboratories found that even at 17 percent solar penetration, integration costs in Arizona were US$4.4 per megawatt hour (MWh).[29] A recent survey of the literature by the White House Council of Economic Advisors (CEA) that centered on wind power found integration costs to be less than US$6/MWh. The CEA report also found that there was no correlation between levels of variable renewable penetration and ancillary service costs.[30] A PNNL study of NG Energy's southern Nevada system found that the annual cost of integrating 150 MW to 1,000 MW of large PV and distributed generation capacity ranged from US$3 to US$8 per MWh.[31] In the United Kingdom, Vivid Economics estimated renewable system integration cost to be around £10/MWh at low penetration levels.[32] Based on this literature, Brown et al. (2019) estimate that an upper limit (and conservative estimate) of system integration costs for solar and wind power would be approximately US$10/MWh, representing a premium of about 5 to 20 percent. Such premiums are too small for CCS proponents to claim that their solution is commercially viable.[33]

Nevertheless, variable renewable energy systems require baseload backup at high levels of penetration to ensure system reliability and resilience; and this comes at a cost.[34] Despite the irrationality inherent in the CCS sweep-it-under-the-rug solution, the uncertain costs of renewable energy storage at high levels of penetration give rise to market uncertainty, empowering CCS advocates. CCS is currently the best option that fossil fuel interests have for preserving the status quo for as long as possible. Its mere existence provides hope to budget-constrained policymakers that there might be another solution to engendering a low-carbon energy future—a solution that does not involve an expensive and disruptive reengineering of the global energy infrastructure.

If the actual rate of fossil fuel price escalation mirrors the estimates in the Energy Information Administration (EIA)'s "new policies" or "current policies" scenario outlined in chapter 1, it is likely that the question of adopting

CCS technology will be moot because the inflated cost of fossil fuel stocks will make coal-fired or gas-fired power plants prohibitively expensive to operate. Under the *new policies scenario*, compared to 2017 price levels, by 2040 the price of oil will rise from US$52/barrel to US$112/barrel, the price of European Union (EU) natural gas will rise from US$5.8 per million Btu (MBtu) to US$9.0/MBtu, and the price of U.S. coal will rise from US$60/tonne to $64/tonne (refer back to table 1.4). Similarly, under the *current policies scenario*, compared to 2017 price levels, by 2040 the price of oil will increase 160 percent to US$137/barrel, the price of EU natural gas will rise 62 percent to US$9.4/MBtu, and the price of coal will increase 15 percent to US$69/tonne. Indeed, as we saw in chapter 1, some estimates already indicate that solar PV and onshore wind are less costly than coal-fired power even without CCS technology added (refer back to table 1.8).[35] Accordingly, each new technological advance in the renewable energy space represents yet another spadefull of dirt on the grave of this illogical technological response.

However, indications suggest that this source of uncertainty—resistance from incumbent market leaders—will likely be far less influential by 2025; in the interim, there are a number of powerful stakeholders who will extol the wonders of this kamikaze solution because there is a lot of sunken investment that still needs to be amortized. In chapter 4, we will examine strategies to attenuate such resistance by explaining strategies to better manage technological innovation and the uncertainties that confound identification of optimal energy transition paths. Chapter 4 paints an energy sector vista that is inherently more empowering (forgive the pun) because it highlights trends toward more decentralized control of energy systems that permit a far broader array of stakeholders to profit from the transition.

DIFFICULTIES IN FOSTERING AND FINANCING ENERGY INFRASTRUCTURE TRANSITIONS

Regardless of where one stands on the topic of what constitutes a prudent target for greenhouse gas emissions, one must acknowledge that a disparity exists between what is ecologically or scientifically desirable and what is politically feasible. As we tell our students, the physicist would say it's prudent to scrap investment in a coal-fired power plant when its energy payback ratio is below 1—that is, it takes more energy to extract than one can get out of it. The climate change scientist would argue that the plant should be shuttered immediately. The economist would say it's prudent to scrap investment when its cost-benefit ratio falls below 1, that is, when it costs more to produce than one can make money out of it. However, many politicians would say it's only prudent to scrap an investment when there are no jobs at stake—a situation that will *never* occur at an operational facility; or when there is spare financing to support change—another rarity.[36] In this context, energy economics plays a gatekeeping role in influencing a nation's capacity and willingness to finance a transition.

To plan for the future, national energy planners must begin any strategic assessment with an analysis of the energy systems that currently exist, for two reasons: the profiles of existing energy systems largely frame the speed with which a nation can incorporate renewable energy flows into existing infrastructure, and the age of existing energy systems frames the financial ability of a nation to facilitate expedient change.[37]

Consider the first factor—the nature of existing energy systems—and how this impacts national capacity to accommodate a technological transition. On one hand, a nation like Canada, for example, generates approximately 60 percent of its energy through hydropower—a highly responsive energy source. Therefore, if Canada's grids were well connected, the nation would be well equipped to integrate stochastic renewable energy flows into its grids.[38] On the other hand, a nation like China generates about 65 percent of its electricity through coal-fired power plants (baseload systems that cannot be expeditiously adjusted to complement stochastic forms of energy). Therefore, China is not well equipped to balance large contributions from energy technologies such as wind power or solar PV.[39] In both nations, the requisite expenditure for a transition will be enormous; but in China's case, an expansion into renewable energy requires not only an investment in new plants and transmission and distribution (T&D) networks, but also the addition of more peak-load power generation capacity.

The second structural factor that moderates the prospects of a rapid transition to renewable energy in a given nation relates to the age of existing energy infrastructure and market structure. Energy plants, like all major capital expenditures, are subject to depreciation. Energy providers that manage these systems have developed operational budgets that include depreciation expenses. For example, an energy provider that erects a coal-fired power plant might depreciate the plant over a thirty-year period, meaning that each year one-thirtieth of the investment cost will be written off as an operational expense. When a plant is shuttered, any amount that has not been depreciated has to be listed in that year as an expense. Due to the high capital outlay of utility-scale power plants, replacing undepreciated assets with new renewable capacity can significantly impair the financial health of an energy provider. Consequently, energy providers—whether they are private-sector firms or national utilities—are rarely keen to prematurely shut down polluting power plants in order to reduce greenhouse gas emissions. This problem suggests that nations such as the United States that are saddled with fleets of older energy plants are able to financially implement a transition at a faster pace than nations such as China that have been building new coal-fired power plants at breakneck speed over the past two decades.[40]

Chapter 5 will examine this challenge of fostering and financing the energy infrastructure transition. Due to the scale and scope of investment required, this transition will benefit from international cooperation in research and development and local cooperation in implementation. Policy lies at the core of unlocking financial access to foster change; and this means that a better understanding of policy interventions is needed.

DIFFICULTIES IN GETTING THE POLICY MIX RIGHT TO ENGENDER MARKET CONFIDENCE AND SUPPORT INNOVATION AND TRANSITION

Energy policy can be conceptualized as a problem-solving process through which undesirable behavior is modified and markets are redefined to achieve specific outcomes. When only small modifications are needed to solve problems, policy adjustments tend to be of a gradual nature. However, when sweeping change is required, policy strategies tend to reflect the complexity of the markets into which they are applied. In short, complex change begets complex policy interventions.

Within this backdrop, it is possible to conceptualize technological innovation as falling into one of two forms—linear or disruptive. Linear innovation builds on existing technology. For example, the electric typewriter retained many of the features of the manual typewriter. On the other hand, disruptive technology departs from existing technology in a manner that is significant enough to disrupt markets, dislodging market leaders and nurturing the emergence of new technological champions.[41] For example, when the computer usurped the electric typewriter as the preferred mode of written office communication, IBM and, to a lesser extent, Hewlett-Packard, were the only major typewriter manufacturers to make the transition. Similarly, as mobile phone technology transitioned to smart phones that stressed computing power over connectivity, market leaders such a Nokia and Blackberry slid into oblivion.

When a sector's technology is developing linearly, market evolution is more predictable because evolving cost profiles and changes in demand can be somewhat extrapolated from historical precedent. In most industries, though, the pursuit of progressive innovation inevitably leads firms to become increasingly out of touch with market needs—opening doors of opportunity for new technologies to challenge incumbents for market share.[42] Evolving market needs yield uncertainty because new technologies, which fulfill needs in unexpected ways (e.g., by amplifying demand), can emerge and sport cost profiles that cannot be predicted from historical trends, because there are no historical trends. The energy sector is currently host to a number of disruptive technologies, making it difficult to predict a winner. It infuses an otherwise certain event (a transition away from fossil fuel technologies) with a high degree of uncertainty (which technology will dominate) and complicates the policymaking process. Let us consider a few areas of contention and the resulting policy complications.

For starters, the future of nuclear power is hotly contested. If one were to believe some nuclear supporters, the low-carbon profile of nuclear power could catalyze a renaissance in capacity expansion.[43] However, the promise of nuclear power, as embraced by proponents, does not reflect current trends. Globally, the amount of nuclear power generated, as measured in terawatt hours, has actually declined since 2005.[44] The decision in Japan to shut down the reactor fleet post-Fukushima pending safety checks has only exacerbated this decline. For the nuclear power industry, Asia currently serves as the only region with substantial growth potential, with China in particular committed to nuclear power plant expansion. According to the World Nuclear Association, as of 2018, there were 130 operable nuclear power reactors in Asia, 35 under construction, and commitments to build a further 70 to 80 plants in upcoming years.[45] In merits reiterating, though, that market potential in Asia might be curtailed somewhat due to a policy change by the Moon Jae-in administration in South Korea, which appears intent on curbing nuclear power development, ironically despite public support. Disconcertingly, one of the authors of this book (Sovacool) has argued that the cost associated with producing safe, reliable, nuclear power (advanced nuclear reactors, secure permanent storage, responsible decommission plans) is so significant that nations have only two strategies at hand: massively subsidize costs so it is safe, or cut corners and run the risks of disaster to financially justify the commitment.[46] Nuclear power can be safe, or cheap, but not both.

Since the 1950s, the promise that nuclear power will be "too cheap to meter" has shrouded the industry in an intoxicating haze that some technocratic nations find difficult to resist.[47] According to the World Nuclear Association, there are currently 20 nuclear power reactors under construction in China alone, with another 41 units planned.[48] As scores of reactors come online in China over the next two decades, a curious geopolitical dynamic will develop in Asia. The risks posed by these reactors will in many ways nullify any additional risks that neighboring nations might undertake by incorporating contributions from nuclear power. After all, what additional risk does the restart of Japan's reactors pose to Japanese citizens when across the China Sea, Chinese reactors are springing up with far less regulatory oversight? In the absence of another major nuclear disaster, it remains to be seen how many other nations embrace this "well, if my neighbor's doing it, I might as well do it too" mindset. According to the International Atomic Energy Agency, approximately sixty countries have expressed interest in initiating nuclear power programs.[49]

In direct competition to a nuclear renaissance, there are hosts of renewable energy technologies that are currently commercially viable. We talked about many of these in chapter 1, especially wind and solar, which can be cheaper than fossil fuels already under certain circumstances, and the nature of these circumstances radically alters the viability of supporting one technology over another.

The phrase "under certain circumstances" suggests that a number of factors influence the economics of renewable energy technologies. For example, Australia boasts some of the largest geothermal energy potential in the world; however, these geothermal sources are located far from Australia's major demand centers, elevating the cost of exploiting an otherwise cheap form of energy.[50] Conversely, Iceland's geothermal resources are close to the main city of Reykjavík and, as a result, deliver cheap, clean energy to its citizens. In China, much of the nation's wind power potential lies in the sparsely populated northern regions, amplifying T&D costs.[51] Although the same is true of Germany's wind power potential (high potential in its more rural northern regions), the distances that must be traversed to distribute wind power to Germany's demand centers are far shorter.[52] Overall, geography, population densities and distribution, climatic characteristics, the cost of land, the cost of labor, and levels of social acceptance can all play a decisive role in dictating whether a given renewable energy technology will be capable of generating energy at optimal cost, and these factors influence the nature of policy support.

Another factor that complicates predictions of which emergent renewable technology will rule the roost is the observation that the innovation trajectories that apply to each of these technologies are not the same. Some technologies will mature faster than others. Of specific relevance, the manner in which the cost profiles of these new technologies respond to economies of scale and new innovation is very difficult, if not impossible, to predict.

Take, for example, wind power—one of the more mature renewable energy technologies. A report by the IEA contends that, at a minimum, wind power costs are likely to decline by 1 percent per year between 2012 and 2030, with a probable aggregate cost reduction of between 20 to 30 percent by 2030.[53] Since wind power is a mature technology, it seems intuitively reasonable that slow but steady cost improvements will characterize industry development going forward. In reality, though, the only thing that is mature about this industry is the evolution of vertical axis wind systems. Other designs could revolutionize wind power development. For example, researchers at the University of Kyushu have developed a wind lens that is essentially a shroud that fits over a wind turbine blade. Under experimental conditions, this shroud allows a wind system to capture between 2.5 and 2.8 times the amount of wind energy that a traditional wind turbine can capture.[54] If the barriers that bar commercialization of this technology can be overcome, this is the type of innovation that can significantly sway the economic cost trajectory of offshore wind power.

Similarly, the evolving cost profile of solar PV is highly unpredictable. The years between 2010 and 2014 represented a watershed period for solar energy, as the cost of solar PV cells declined precipitously, thrusting solar PV onto the agendas of energy planners. Table 2.1 shows just how fast costs fell. In the five-year period, the weighted average installed cost of solar PV decreased 39 to 58 percent; and as a result, cumulative installed capacity grew over fourfold. Policymakers need to know how much further solar PV costs will plummet in order to define a place for this technology in national energy mixes.

Predicting costs in the solar PV sector is not an easy task. Since 2015, the solar PV market has burgeoned as costs continue to decline and new investment pours into the sector. Over the last two years alone, global installed

TABLE 2.1 Performance Statistics in the Solar PV Industry

	2010	2013	2014	2010–2014 (% change)
New capacity additions (GW)	16	39	40+	150+
Cumulative installed capacity (GW)	39	139	179+	360+
Regional weighted average installed cost, utility-scale (2014 US$/kW)	3,700–7,060	1,690–4,250	1,570–4,340	−39 to −58
Regional weighted average utility-scale LCOE* (2014 US$/kWh)	0.23–0.5	0.12–0.24	0.11–0.28	−44 to −52
Residential LCOE in selected countries (2014 US$/kWh)	0.33–0.92	0.15–0.49	0.14–0.47	−49 to −58

Source: Renewable Energy Policy Network for the 21st Century (REN21), *Renewables 2018 Global Status* Report, http://www.ren21.net/gsr-2018/; International Renewable Energy Agency (IRENA), *Renewable Power Generation Costs in 2014* (Bonn, Germany: IRENA, 2015).

*LCOE = levelized cost of electricity.

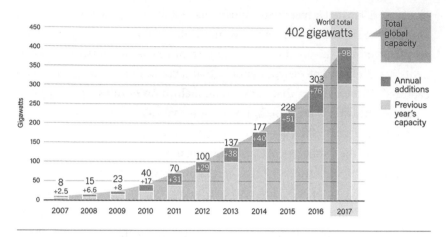

FIGURE 2.3 Solar PV global capacity and annual additions, 2007–2017 (Renewable Energy Policy Network for the 21st Century [REN21], *Renewables 2018 Global Status Report*, http://www.ren21.net/gsr-2018/.)

capacity has increased by 76 percent (figure 2.3). Compared to coal-fired power, solar PV is now, in many situations, commercially competitive.[55]

Yet, even as prices stabilize at commercial levels, within the solar PV sector, the promise of disruptive technological advances further complicates market analysis. Take, for example, a translucent solar photovoltaic cell that was developed by researchers at MIT. In 2011, the inventors published a paper that claimed 1 to 2 percent conversion efficiency.[56] A mere four years later, a firm founded by these pioneers (Ubiquitous Energy) claimed to have a product that is 90 percent transparent and capable of 10 percent energy conversion efficiency.[57] Researchers at the University of California Los Angeles boast similar promising results with a different technological platform featuring 70 percent transparent cells that can produce energy at 7.3 percent energy conversion efficiency.[58] Even at 7 to 10 percent efficiency, the applications for solar PV film technology in an urban setting or in transport renders this to be a product evolution that could significantly alter the fortunes of solar PV firms in the highly competitive utility-scale marketplace. Moreover, as the cost of these new technologies drops in response to economies of scale and improved research and development, the potential of these technologies to contribute to spillover production cost improvements of traditional solar PV cells is high.

Spillovers of this type remind us that the uncertainty of success that shrouds new technologies depends significantly on interconnections between technologies. For example, if smart electricity grids and electric vehicles emerge as market leaders in the new energy age, the potential for renewable energy technologies that are plagued by stochastic power production profiles (i.e., wind power, solar power, etc.) will increase exponentially because both smart grids and electric vehicles will be able to store some of the surplus energy that these stochastic technologies generate. The importance of exploiting technological synergies was at the core of Tesla's decision to fund the establishment of an energy storage firm—Tesla Energy—and to purchase one of the United States's largest solar PV manufacturers, SolarCity, in 2016, creating "the world's only integrated sustainable energy company."[59]

Market uncertainty wrought by new technologies differs markedly from market uncertainty caused by the type of fossil fuel industry resistance that was discussed earlier. The uncertainty associated with challenges from fossil fuel firms—via erecting political obstacles and advancing the interests of CCS technology—will likely only prevail for another decade or so of competitive pressures. It will pass. Conversely, the uncertainty posed by an ever-evolving host of new technologies will be a defining feature of the competitive landscape in the energy field for decades to come. As more investment moves into renewable energy, new commercially viable technologies will emerge, and some will establish leadership positions (leveraging economies of scale to elevate research and development and widen the competitive divide). They will then become the entrenched, undisputed market leaders, and these technologies will assume the moniker of "conventional energy technologies." When this happens, the cycle will have traveled full circle and market predictions will become more certain until new disruptive technologies emerge. For now, we must face a period of market instability caused by competing disruptive technologies. Predicting winners and losers will be difficult, making it hard for policymakers to champion one technology over another.

With this challenge in mind, chapter 6 will examine the array of policy instruments that are being applied around the world to ensure that technological competition is maintained and that forces to incentivize a transition continue to yield results even as market dynamics evolve. Yet the fate of new technology represents only the supply side of the equation when it comes to market uncertainty. Over on the other side of the ledger—the demand side—there are different uncertainties that will also require policy intervention. As such, the focus on policy instruments in chapter 6 will also have relevance for chapter 7, which centers on consumer apathy.

CONSUMER APATHY

Humanity has sadly become addicted to the unrelenting pursuit of the acquisition of stuff, seemingly as a proxy for the pursuit of happiness.[60] This gives rise to a two-headed problem when it comes to enabling an energy transition. First, the modern worldview is of consumers serving the economy, not the other way around[61]—creating more stuff to feed the insatiable appetites of consumers amplifies energy demand. So not only are we confronting a need to transition out of fossil fuel–based energy forms, but we also need to add ever-more electricity generation capacity to the grid to support the supersizing of economic activity. Second, on an individuallevel, those who live in developed nations are hard-pressed to dial back consumption to sustainable levels, even though they are sympathetic to environmental causes. Even the energy usage profiles of many prominent environmentalists from developed nations yield shockingly unsustainable energy consumption levels if applied to all inhabitants of earth. There is good reason for this: such people are in great demand and they travel frequently. However, justifications aside, this point illustrates just how difficult it is to rein in consumption to live a truly sustainable energy existence. If dedicated environmentalists find this challenging, how about the rest of us?

Consider for a moment the habits that you might have developed as an energy consumer. If you are like most people, you likely have a number of appliances in your home that are currently plugged in and drawing energy—as they sit on standby. You might use energy for many nonessential uses, such as powering aromatherapy diffusers, lighting your garden at night, providing mood music in certain areas of the home, warming toilet seats, or popping microwave popcorn. Human beings are, by and large, oblivious to the multitude of ways in which they exacerbate energy demand through innocuous yet energy-consuming day-to-day activities. These little niceties have become so entrenched in our lives that removing them might require some serious ideological surgery.

One of the reasons that we have become so desensitized to the role that our behavior plays in shaping energy demand is that we are, by and large, rarely able to witness the full impact that these activities have on our environment. For starters, in many advanced economies, the most polluting industries have moved overseas to developing nations. We have essentially exported our pollution to the Chinas and Indias of the world—as the expression goes, *out of sight, out of mind*.[62] In a similar vein, the numerous fossil fuel power plants that do exist in most advanced economies chug away in locales that very few of us frequent; and those who do traverse such areas rarely give full attention to the environmental significance of a smokestack that appears to be emitting a steady stream of white smoke. Finally, when it comes to the threats posed by climate change, many citizens in advanced economies don't understand the enormity of the threat because the worst is yet to materialize. Certainly the odd storm hits a coastal area in the United States or a flood washes out a few roads in a village in the United Kingdom; however, these damaging events are infrequent and often not easy to connect to climate change. In short, we just do not see the damage that our consumptive excesses are causing.

Part of this lack of understanding stems ironically from our ultraconnected world. We are bombarded by information and face unprecedented time demands. We live in a multitasking age where downtime is rare. Therefore, there is less time for thinking through the implications of societal trends; and even when there is time to think, well maybe we had better check our e-mails first. Indeed, an argument can be made that most people do not want to think about such things; they simply want to be told what the problem is and assured that someone is addressing it. As the sociologist Paul Edwards once remarked, "The most salient characteristic of technology in the modern (industrial and postindustrial) world is the degree to which most technology is not salient for most people, most of the time."[63]

These three forces—excessive materialistic pursuits, ignorance of the impacts, and lack of time to become informed—have sired societies that have become divided by polarized worldviews in many areas, including energy. On one side, we find liberal thinkers who point to increasing levels of greenhouse gas emissions and alarmingly high levels of premature deaths associated with particulate matter inhalation, all the while admonishing human beings to rethink what it means to lead a happy, sustainable existence. On the other side, we find conservative thinkers who are largely risk-averse and unable (or unwilling) to make the connections between their energy choices and global environmental problems. This latter group is often comforted by media sources that tell them that everything is all right—sources nurtured by special-interest groups who are well funded by fossil fuel energy firms.

There is, of course, good reason for energy special-interest groups to strive to influence consumer opinion. Almost $1 in every $7 to $10 eventually makes its way to an energy-related firm, whether this is directly through energy purchases or indirectly through the purchase of consumer goods that have energy costs embedded into their prices.[64] There are trillions of dollars invested in fossil fuel infrastructure around the world. As mentioned earlier, for many fossil fuel energy firms, even if they wanted to switch to cleaner technologies, the prospects of doing so in an expedient fashion are inhibited by the need to fully depreciate capital investments.

With that said, energy firms are not the only entities that have good reason to resist a mandate for transitioning to cleaner energy systems. Consider the average person on the street and the investments that he or she has made in technologies that lock in behavior. In advanced nations, most people own a car or motorcycle that is powered by an internal combustion engine. The switching costs here are high. Most people also likely live in residences that are lit by electricity infrastructure that is incapable of delivering energy back to the grid. To move to a residential smart grid would require costly reengineering of the average home's electricity systems. Most people also possess a number of electrical devices that were purchased in the past and are no longer the most energy-efficient models on the market—if they ever were in the first place. Once again, a shift to more efficient solutions requires both time and money. In short, for many consumers, the notion of an energy transition might have an altruistic appeal, but in practice, it is an initiative that does not deliver immediate relative advantages.[65]

If a widespread transition away from carbon-intensive energy technologies is to be expedited, a portfolio of responses needs to be developed to reverse such consumer apathy. Failure to do so will result in stores full of dusty new technologies, and that is never a good recipe for developing sustainable markets. Chapter 7 picks up on this challenge and analyzes the problem in far more detail so we can identify targeted solutions to this conundrum. We consumers may just need to lead this change because, as the next section discusses, political change might just be the biggest challenge of them all.

FAILURES IN POLITICS AND GOVERNANCE

Irrespective of the type of political system, politicians live in a world of short-term decision making. In uncontested regimes such as the one found in China, politicians look to maximize short-term gain and distribute benefits in order to avoid the emergence of public dissent.[66] In democratic regimes, the mechanisms underpinning short-termism are similar. The fortunes of politicians (potential of reelection) depend extensively on voters' perceptions of a governing regime's performance or potential performance. In practice, this situation engenders a mindset in which governing politicians are motivated to fast-track benefits to ensure that they accrue during politicians' terms in office and to postpone costs to ensure that they materialize after politicians' terms in office. Short-termism yields dysfunctional policies when it comes to climate change mitigation and energy transition, because costs need to be incurred now for benefits that will not be reaped for decades.

This is not to say that all politicians are purely self-serving or focused only on extending their tenures in office. Indeed, one can readily see the logic in an argument that "pay now, possibly benefit later" strategies might not serve taxpayers as well as "pay now, benefit now" strategies. Policymakers face difficult financial allocation decisions. Should they commit funds to bolster national healthcare or reinforce social security? Should they open new schools or build more public housing? In many cases, solving these problems means far more to the average voters who demand results from their tax dollars. In short, any democratically elected government that decides to throw its weight behind climate change mitigation strategies assumes a risk that citizens possess both the foresight to reward long-term planning and the resilience (and selflessness) to put off present benefits for future rewards.

This political barrier is exacerbated by the uncertainty associated with climate change science and by difficulties in estimating the progress of renewable energy technologies (both outlined in previous sections). The logic goes, why trade off funding other important policy areas to invest in comparatively expensive energy transition initiatives when the same transition can be enacted in five years' time, once technological innovation has lowered costs? Of course, the only justification for expedience is if the transition needs to be immediate in order to avert climate change disaster. However, given the uncertainties associated with the mitigation efforts of other nations, elected officials undertake yet another bold risk by curtailing expenditures in other needy areas to finance a transition that might not even help solve the problem. In chapter 8, we explore these types of policy trade-offs in greater detail, consider the challenges inherent in transboundary energy issues, and investigate a number of other energy governance barriers that merit attention in fostering a faster transition.

In conclusion, then, the energy sector provides a rich environment in which to explore the dynamics between the certainty that exists when a problem clearly needs to be addressed and the uncertainties inherent in deciding among competing solutions. As was discussed, competitive resistance from powerful incumbent market leaders and uncertainties associated with technological evolution confound the identification of clear transitional certainties. However, even if the desired destination were clear, political will must exist to support top-down expedient change. In this regard, scientific dissent over the perils of climate change, the composition and age of national energy infrastructure, and political short-termism have sired conditions that can make it very difficult for politicians to get behind a costly transition plan. In other words, contrary to the old adage "where there is a will there is a way," in energy both the way and the will are lacking.

FAST-TRACKING TRANSITIONS

In the first chapter, ten trends were presented that point toward an inevitable transition away from carbon-intensive energy systems. In this second chapter, factors that render the scale, scope, and pace of this certain transition less predictable were introduced. In a nutshell, we saw that market uncertainty arises from two factors: (1) attempts by powerful vested interests to preserve the status quo and (2) lack of clarity regarding which competing technology will ascend to market leadership amid this transition. The pace of diffusion is also heavily influenced by scientific dissent regarding climate change, the composition and age of national energy systems, and the short-termism associated with policymaking. In aggregate, these sources of uncertainty have delivered us to the precipice of a conundrum: it is imperative that the transition away from carbon-intensive energy systems be expedited to avoid the worst perils attributed to climate change, yet barriers to diffusion appear intractable.

We believe that there are solutions, albeit potentially thorny, and these will be outlined in the concluding chapter, chapter 9. There is a lot at stake in this transition. Fortunes will be gained and lost. Jobs will be created and destroyed. Moreover, the state in which we leave the planet for future generations will be significantly impacted by the decisions we make today (and to a degree tomorrow). Species of flora and fauna that have survived centuries of natural calamities will not be around at the end of all of this. For many animals that we share this planet with, we have already cast the dice against their survival.

In many ways, with or without an energy transition, our lives will be different. Approaches to how we undertake our daily activities, how businesses conceptualize operations, and how governments, businesses, and citizens interact will need to change—regardless of the path we end up taking.

It is with all this in mind that we decided to write this book. If given the choice, we choose the path that rewards courage and tempers the discomfort of change with a desire to foster progress.

In this regard, the next seven chapters represent a diagnosis and prescription of sorts. The next six chapters will explore the challenges introduced in this chapter in far greater detail. The intent is to identify the causal relationships underpinning the challenges in fostering an expedient energy transition and then use these insights to put forward recommendations that can help foster a faster transition that empowers and enriches in the medium to long term. The final chapter presents our prescription—one that we hope is broad enough to be of relevance to policymakers and stakeholders in any nation that wish to play a role in effecting change. We are writing about a topic of grave consequence and peril, but this book is really about opportunity and hope.

3

The Uncertainties
of Climate Change

I n the leadup to and the aftermath of the launch of the Kyoto
Protocol in 1997, there was a great deal of media coverage centering on fundamental questions such as "Is climate change a
valid threat to humanity; and if so, is it true that human activity
is to blame for the perils to come?" During those days of climate
skepticism, however, we have witnessed the unfortunate proliferation of denial that seems to have been spawned by a mix of incomplete information, ignorance, special-interest influence, and gin.

In this chapter, we review some of the prevalent misconcep-
tions that skeptics have trotted out in their attacks on science.
This review serves to document the creative strategies employed
by special interests to undermine support for climate change
mitigation efforts, but it also highlights how far we have come
in clarifying key issues centered on climate science. The chapter then turns to a review of climate change projections to get
a better sense of why many scientists consider climate change
mitigation efforts to be so urgent. We then examine the vested
interests that fuel opposition to change, explain why these
stakeholders view emerging scientific projections as threats,
and outline the problems that arise when the rich and powerful

resist change. Finally, the chapter concludes with a discussion that centers on the Paris Agreement, an agreement that largely mirrors the gradual erosion of special-interest opposition in the face of forces that increasingly compel us to act.

OF VOLCANOES, SUNSPOTS, AND FRIENDS:
A SHORT HISTORY OF SKEPTICISM AND DENIAL

Although lacking in rigor and perhaps integrity, climate change skeptics do have one thing going for them: creativity. In the beginning, skeptics flat out denied that the world was even warming. They would point to record snow storms in Topeka, Kansas, or unseasonable cold spells in France, providing anecdotal evidence to argue that, if the world is indeed warming, it sure doesn't explain all the snow shoveling going on. For example, an article in the United Kingdom's *Daily Mail* declared that a record-setting cold spell in Montana and a historical record for early snowfall in Austria provided evidence that climate skeptics have a point.[1]

This argument contributed to concerted efforts to reposition this phenomenon as "climate change" rather than "global warming." Although it is accurate to say that our atmosphere is retaining more heat because of greenhouse gas (GHG) accumulation, this does not mean that all places on the planet are affected in the same way. As the atmosphere heats, its capacity to hold water increases. At the same time, water in oceans and lakes warms and evaporates at higher rates, elevating atmospheric water content. As oceans warm, even imperceptibly for the average swimmer, oceanic currents also begin to change because of changes in water temperature gradients—and as a result, air circulatory patterns shift. In other words, it is entirely expected among atmospheric scientists that a warming planet will give rise to regions where rainfall patterns and snow might actually increase.[2] In a nutshell, some places on the planet will heat up while others might cool down; this explains why *climate change* is perhaps a better moniker for the phenomenon that we are experiencing. However, there should be no mistake regarding the trends we are observing; in aggregate, global temperatures are rising and will continue to rise as GHG molecules in our

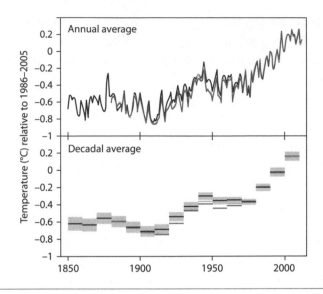

FIGURE 3.1 Observed globally averaged combined land and ocean surface temperature anomaly, 1850–2012 (*Climate Change 2014: Synthesis Report. Contribution of Working Groups I, II and III to the Fifth Assessment Report of the Intergovernmental Panel on Climate Change*, ed. R. K. Pachauri and L. A. Meyer [Geneva, Switzerland: IPCC, 2014].)

atmosphere absorb heat. Figure 3.1 shows the overall global temperature rise over the past 150+ years.[3]

The second grouping of creative climate skepticism came veiled in attempts to explain away human contributions to this problem by demonstrating that the warming is either cyclical or of an episodic nature. According to this line of reasoning, the culprit behind global warming trends was not mankind but rather Mother Nature, bent on yet another irrational tirade.

Some argued in the 1990s that the warming trend that scientists were observing was the result of an overactive hydrogen reactor that we unfortunately depend on for heat—our sun. Skeptics argued that our sun occasionally belches out swirling masses of heat, commonly known as solar flares. These solar flares roll across the universe like thermal tsunamis, reaching us on earth a few days after they are emitted from the sun. The fundamental science behind such a supposition is true; the sun does emit solar flares, and they can indeed cause temporary warming effects (and mess with our

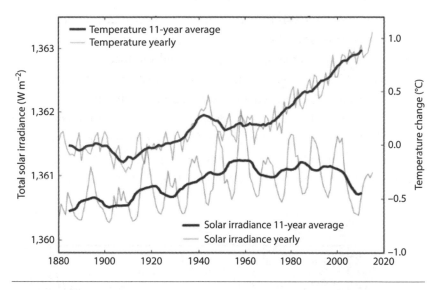

FIGURE 3.2 Sunspot and temperature records (Taken from skepticalscience. com based on temperature data from NASA Goddard Institute for Space Studies; 1880–1978 annual total solar irradiance [TSI] data from Krivova et al., 2007; 1979–2015 TSI data from the World Radiation Center.)

electronic systems) when they eventually reach the earth. In a 2004 article, the BBC reported speculations by a scientist, Dr. Sami Solanki, who, after observing the record of sunspots (which produce solar flares) over the last 1,115 years, concluded that in the past sixty years sunspot activity has been far more frequent. He said that this might be a source of global warming.[4]

Unfortunately, the researchers missed one small detail: the sunspot activity had tailed off over the past twenty-five years or so. In other words, over a period when sunspot events decreased, the planet continued to warm. Figure 3.2 plots this relationship.[5]

Taking another line of attack, some skeptics began to champion a theory centering on volcanic activity. The postulation was that GHG emissions from volcanoes are far greater than human GHG emissions are; and they are very difficult to measure because active volcanoes tend to be . . . hot. This makes it hard to accurately measure GHG emissions from volcanoes. As one newspaper report speculated in 2009, "Over the past 250 years, humans have added just one part of CO_2 in 10,000 to the atmosphere. One volcanic cough can do this in a day."[6] This was couched in scientific truth—volcanoes are responsible for discharges of GHG emissions (albeit not to the extent claimed in this newspaper report)—and the theory that volcanoes are largely to blame for global warming was virtually untestable. Or was it? In 2004, a group of researchers from the National Center for Atmospheric Research conducted a number of tests to try to evaluate the correlation between solar activity, volcanic activity, and the temperature record.[7] The charts that documented the findings have been reproduced in figure 3.3. Clearly, neither of these factors explains the temperature increase.

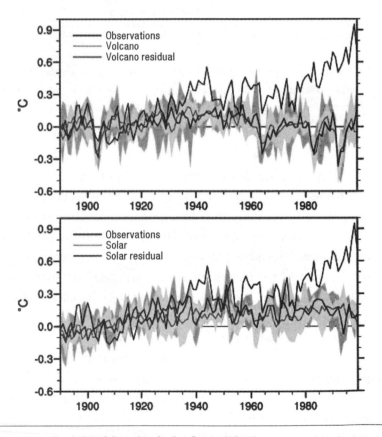

FIGURE 3.3 Analysis of the role of solar flares and volcanic emissions in temperature increase (G. A. Meehl et al., "Combinations of Natural and Anthropogenic Forcings in Twentieth-Century Climate," *Journal of Climate* 17 [2004]: 3721–3727.)

In the same analysis the scientists tested the link between all natural sources of GHG, anthropogenic (human-caused) sources only, and then both combined. Figure 3.4 depicts the fit between the trend lines of these factors and the observed temperature record in black. As should be apparent, anthropogenic emissions come closest to explaining the observed temperature record. When natural sources are added in, the fit is almost perfect.[8]

In summary, figure 3.4 tells us that human-induced GHG emissions have predominantly driven the rise in global temperature observed since the 1900s. While it is true that there are also natural influences such as volcanic activity and sunspots, these influences play very minor roles in explaining the sudden escalation of global temperatures.

A final colony of skeptics that has scuttled out of the woodwork can be defined as the "climate change is our friend" group. Some skeptics from this camp argue that for nations closer to the Arctic and Antarctic poles, climate change is going to have a positive impact. In nations like Canada and Russia, the argument goes, the economic benefits will be significant as warmer temperatures will free up more land for economic activities. A highly contested study published in *Nature* agrees with part of this hypothesis, estimating that a minority of nations in cooler countries could experience up to a 20 percent economic gain from climate change by 2100. Indeed, Robert Mendelsohn forecasts that by 2200, parkland and temperate forests could replace tundra and boreal forest in high latitudes, including much of Canada and Siberia. At the same time, savannah would replace tropical forest and deserts would replace grassland and parkland in low latitudes, dramatically expanding the range of the Sahara Desert.[9] The *Nature* study projects that the poorest 40 percent of all countries will experience economic declines of up to 75 percent by 2100, relative to a world without climate change.[10]

Nevertheless, this camp of skeptics was emboldened in 2007 when nine ships managed to navigate through the previously impassable Northwest Passage in Canada's north. Economic bounty in the form of shortened shipping routes and enhanced revenues from amplified exploitation of Arctic resources fueled optimism that global warming might indeed be our friend.[11] Refutation of this theory came from within the economic community itself.

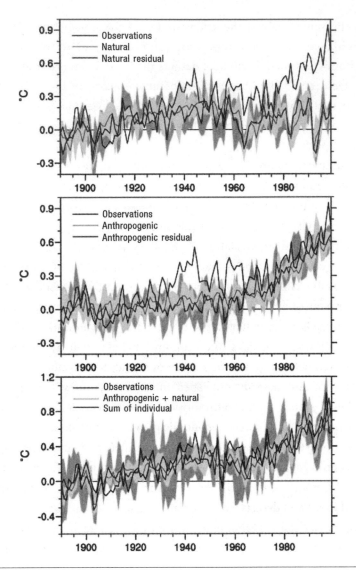

FIGURE 3.4 Natural versus human sources of greenhouse gas emissions and the relationship to the temperature record (G. A. Meehl et al., "Combinations of Natural and Anthropogenic Forcings in Twentieth-Century Climate," *Journal of Climate* 17 [2004]: 3721–3727.)

Although a northern nation such as Canada might derive some benefits from polar warming, studies suggest that other adverse impacts would far eclipse any gains. According to a recent economic analysis by Canada's National Roundtable on the Environment and the Economy, a federal government agency, losses to Canada's economy as a result of climate change are estimated to amount to C$5 billion per year by 2020 and between C$21 and C$43 billion per year by 2050.[12] Evidence of damage is already on display in the Canadian province of British Columbia, where warming in northern latitudes has been linked to wide-scale bark beetle infestation that adversely impacted over 25 million hectares of forest between 2000 and 2010.[13]

Another moonbeam of reason that emerged from the luster and bluster of the "climate change is our friend" campaign centers on the observation that before human beings intervened, the earth was (and still is) entering into a cold phase that will eventually send the earth back to the kind of cold snap last seen during the days of Manny the Mammoth and Sid the sloth. Consequently, the argument goes, a bit of forced warming is in fact helping humanity to delay the adverse impacts of a colder planet. As Dennis Avery, a senior fellow at the Hudson Institute, summarizes, "Two thousand years of published human histories say that the warm periods were good for people. It was the harsh, unstable Dark Ages and Little Ice Age that brought bigger storms, untimely frost, widespread famine and plagues of disease."[14]

The flaw with this assessment centers on a misguided conceptualization of time. Consider what we know will happen. As climate change progresses, the capacity of flora and fauna to adapt will diminish.[15] Moreover, citizens from coastal communities in developing nations where abatement measures are not affordable will be forced to withdraw from their homes because of sea-level rise.[16] It is entirely possible, if not probable, that some of these disenfranchised groups will become mired in conflict because of forced migration or climate change–induced poverty.[17] Then once all the fossils fuels are burned and the sea levels have risen, the global climate will revert to a cooling trajectory, returning us to a path toward an ice age. This is where the time factor comes into play. Transition to an ice age will recommence in about two thousand years on our planet regardless of what we do.[18] Warming our planet now and living with the economic, environmental, and social damage this causes will not significantly alter the planet's long-term course toward a cooling phase. Indeed, in the bigger picture, human-induced climate change is a small blip on the planet's temperature record. But this blip will cause some real damage to people, as well as to the flora and fauna with which we share the planet.

Fortunately for the sake of future generations, scientists have now had the time to research rebuttals and prepare explanations to help skeptics improve their understanding of how climate change will play out outside of the middle school science laboratory in which they concoct their grand theories.[19] Consequently, embarrassed by their myopic assessments, skeptics have virtually disappeared from learned society. This leaves just a few disconcerting voices still barking at the moon.

One is, unfortunately at the time of writing, the president of the United States—Donald Trump—who has famously described global warming as "created by and for the Chinese in order to make U.S. manufacturing non-competitive."[20] If this is true, we should be very concerned, because any nation that can perpetuate such a disaster in such a stealthy manner has powers that would make Superman envious. Trump's initial choice of Environmental Protection Agency administrator, Scott Pruitt, refuses to this day to acknowledge that CO_2 emissions are at the heart of climate change.[21] Fortunately, for many concerned about climate change mitigation, Pruitt's erroneous understanding of climate science was overshadowed by his scandalous use of public financing, leading to his resignation in July 2018. Unfortunately, for many concerned about climate change mitigation, his replacement was Andrew Wheeler, a former energy company lobbyist. In February 2015, climate denier par excellence, Senator James Inhofe, demonstrated that he still didn't grasp the science by famously bringing a snowball into the U.S. Senate, decrying it to be evidence that climate change is a hoax.[22] On the bright side, by the time this book reaches your hands, dear reader, Trump, Wheeler, Inhofe, and the rest of their learning-impaired cronies will hopefully have come and gone like solar flares.

Although most climate change skeptics are now confined to pockets of denial in backwoods regions of the United States and other nations, it is incorrect to conclude that climate scientists are in total agreement. Climate scientists agree that climate change is happening and that human activities are driving this ill-fated bus, but there is still widespread disagreement over how the negative feedbacks will play out under a business-as-usual economic scenario. Basically, scientific projections on climate change impacts range along a broad spectrum with alarming impacts on one end and cataclysmic impacts on the other. It serves to consider a few of the perspectives from both ends of this spectrum.

WHY IPCC PROJECTIONS ARE CONSERVATIVE

When average persons on the street hear a name such as the *Intergovernmental Panel on Climate Change* (IPCC), they might imagine a group of people from scientific fields such as climate science, biology, marine science, geography, engineering, and physics joining up with social scientists, economists, policy academics, nongovernmental organizations, and members of civil society to explore this important issue. If so, they would be correct. This indeed is essentially the makeup of the IPCC: more than 830 authors and review editors from over eighty countries were selected to form the author teams that produced the Fifth Assessment Report (AR5).[23] So it is perhaps natural to place high confidence in the projections that come out of IPCC proceedings.

Unfortunately, two factors in particular partially undermine the accuracy of IPCC projections. First, the science is not—and given current technologies, cannot be—100 percent accurate because this is all taking place within a complex adaptive system where thousands of variables interact. Under such conditions, interpreting what happens if any one influential variable within the system changes hinges on the ability to understand and predict all of the feedback responses that occur in response to such a change. Within the environment under which climate change impact projections are being made, scientific models of climate change impact are significantly undermined by insufficient causal understanding and insufficient capacity to predict human behavior. Let's explore both challenges more closely.

If one wished to evaluate how a block of ice might melt in a glass pitcher as the result of a change in room temperature, one would have to analyze a number of key variables. Let's look at five such variables. First, one would need to know the *size* and *shape of the ice block* to estimate how much of the ice surface might be exposed to this thermal increase and how quickly the heat absorption would radiate inward to the core of the ice block. Second, one would have to measure the *dimensions* of *the glass pitcher* to calculate the rate at which heat from the ambient source (the room temperature) will be absorbed by the water. Third, one would need to know the *temperature of the water* in the pitcher. Fourth, the *thermal retention properties of the pitcher* would have to be understood to estimate how much of the added heat would be retained within the boundaries of the pitcher. Fifth, and critically, one would have to know what the *new room temperature level* would be to estimate the magnitude of thermal increase. Although this is quite a challenge, these variables can be replicated as a controllable experiment, and indeed, the findings of such an experiment can be useful in helping us understand how rapidly ice melts under these conditions.

Now imagine that we want to run the same experiment but we don't know the size and shape of the ice block because the ice block sits in a jar in black ink—so we cannot fully predict the mass below the surface or the internal structure of the ice block (whether it has cracks in it that will allow water to seep in). This is essentially the challenge in estimating polar ice cap characteristics. Further imagine that someone keeps stirring the pitcher with a spoon that is sometimes heated and other times frozen. This is analogous to trying to estimate the impact of marine and wind currents that interact with polar ice sheets. Then imagine that our thermometer was purchased from the dollar store and we are never sure how accurate the readings are. This approximates the risk that is inherent in climate change models. Yet, as we continue to run this fuzzy experiment, we learn more about the properties of the variables that impact ice melt rates. The science is not perfect, but it gets us into a range where ballpark estimates might be quite reliable, if it were not for one more confounding factor—human behavior.

Imagine trying to now conduct the experiment just described with one other twist: a person stops by to randomly adjust the temperature 1°C to 2°C higher every few hours. This replicates the reality that despite global efforts to reduce GHG emissions, emissions continue to progressively increase. Each year, we hear promises from world leaders that policies designed to reduce emissions are in place and will indeed be effective . . . eventually. Each year, progress in this regard falls short.

When uncertainty exists, scientists render projections using ranges. So instead of projecting a 3-meter sea-level rise as a result of a given temperature change, they might estimate sea-level rise to be between 1 and 5 meters with a 95 percent degree of certainty. The problem is that there are colossal economic, environmental, and social implications associated with outcomes on either end of this spectrum. If the sea-level rise amounts to only 1 meter, then some of the damage can be attenuated by using varying degrees of climate-proofing infrastructure and building resilience through community-based adaptation. If sea-level rise is 5 meters, huge tracts of land, including major coastal cities such as New York, Bangkok, and Kolkata, will be rendered uninhabitable. There will be global migrations and trillions of dollars necessarily redirected to adaptation measures. One team of researchers even found that due to climate change, average storm surge damages will likely rise from US$10 to US$40 billion per year in 2014 to possibly US$100 *trillion* by 2100, affecting up to 600 million people.[24] Indeed, by combining future global sea-level rise with tide gauge water levels, another research team expects that today's "once in a century" storm surges might become "once in a decade" storms in the future.[25] When uncertainty is as high as we see with climate change impact projections, projected ranges stretch, making it difficult to determine what type of policy response is needed.

The second factor that influences the accuracy of IPCC climate change impact projections relates to the heavy hand of politics. When average persons see IPCC projections, they understandably tend to look upon them as scientifically accurate—albeit difficult to act on given the disparate impacts under best-case and worse-case scenarios, as well as the diffusion of political responsibility. This is far from the truth. The IPCC process is subject to heavy scrutiny, and this scrutiny tends to result in projections that are highly conservative. For example, for AR5, there were over 2,000 expert reviewers who provided over 140,000 review comments to the 830 authors involved in the process.[26] Under such intense scrutiny, only readily verifiable knowledge is accepted. In the rapidly evolving world of climate science, this fact suggests that many of the newer data are reserved for future reports, once the findings can be sufficiently peer-vetted. As a result, if one were to plot the IPCC risk projections (*likely, highly likely*, etc.) on a chart, the distribution would be significantly skewed to the right—with a long tail that reflects a possibility of dire consequences stemming from outcomes beyond mainstream expectations. For example, consider a prediction that a 2°C warming by 2100 will lead to an average sea-level rise of 1 to 3 meters, with a 95 percent degree of certainty. Now consider what the other 5 percent of uncertainty might constitute. The best-case scenario might be a sealevel rise of 0.5 meter; on the other hand, a worst-case scenario might be a sea-level rise of 6 to 8 meters.

In a nutshell, then, when it comes to IPCC climate change projections, the possibility that actual outcomes will exceed projections is very high. This then begs the question, just what has been projected?

Since 1990, the IPCC has published assessment reports every five to six years. As each new report is released, the ranges for the projections have narrowed to reflect growing scientific certainty over the feedbacks that are occurring. Table 3.1 summarizes the evolution of scientific understanding

TABLE 3.1 The subtle evolution of IPCC assessment reports

AR1: Emissions resulting from human activities are substantially increasing the atmospheric concentrations of greenhouse gases: carbon dioxide, methane, chlorofluorocarbons, and nitrous oxide.

AR5: Anthropogenic greenhouse gas emissions have increased since the preindustrial era, driven largely by economic and population growth, and are now higher than ever. . . . Their effects, together with those of other anthropogenic drivers, have been detected throughout the climate system and are *extremely likely* to have been the dominant cause of the observed warming since the mid-twentieth century.

AR1: The evidence from the modeling studies, observations, and the sensitivity analyses indicates that the sensitivity of global mean surface temperature to doubling of CO_2 is unlikely to lie outside the range of 1.5°C to 4.5°C.

AR5: The increase of global mean surface temperature by the end of the twenty-first century (2081–2100) relative to 1986–2005 is likely 1.4°C to 3.1°C under RCP*6.0 (which is the closest scenario to a doubling of CO_2).

AR1: Global mean surface air temperature has increased by 0.3°C to 0.6°C over the last 100 years.

AR5: The globally averaged combined land and ocean surface temperature data, as calculated by a linear trend, show a warming of 0.85 [0.65 to 1.06] °C over the period 1880–2012.

Source: IPCC website: https://www.ipcc.ch/.
*RCP = Representative Concentration Pathway.

within the IPCC process from the first assessment report in 1990 to the fifth report in 2014. The reader is reminded that, while dire, reality will most likely be much, much worse than these projections.

TIPPING POINTS BEYOND THE IPCC: PERMAFROST, ICE SHELVES, AND THERMOHALENE CIRCULATION

Earlier, we emphasized that the IPCC process was a conservative process that likely underestimates the real perils associated with each added degree of atmospheric warming. There are three risks that merit special attention because any of these three outcomes could radically amplify the damage wrought by climate change: (1) amplification in the release of greenhouse gases stored in permafrost, (2) escalation in the breakup of Antarctic ice shelves, and (3) destabilization of the thermohaline circulation.

A researcher at NASA has estimated that permafrost soils have, over the years, sequestered between 1,400 and 1,850 petagrams of organic carbon. To put this into context, it has been estimated that carbon injected into the atmosphere from all anthropogenic sources since 1850 amounts to about 350 petagrams. Ominously, the IPCC estimates that the area of permafrost near the surface (upper 3.5 meters) is likely to decrease by 37 percent (RCP2.6) to 81 percent (RCP8.5) for the multimodel average (medium confidence).[27] Yet perspectives are mixed. At one extreme, some contend that most of the carbon that has been sequestered in permafrost areas is near the surface and so a warming of the Arctic (in particular) could trigger a massive release of methane and CO_2.[28] At the other extreme, a recent study looking specifically into methane emissions concluded that the emissions might not be as high as previous studies that informed the IPCC process suggested.[29] Estimating how much of the trapped GHG will be released as permafrost melts is an imperfect process. Melt rates are influenced by confounding factors such as albedo rates, contact with water (or rain), depth of the permafrost or ice under investigation, and climatic changes. Undoubtedly, we are going to experience some added emissions irrespective of what we do. However, if the worst fears are realized on this front, we might find ourselves facing a runaway warming process.

The second threat involves wide-scale melting of the Greenland and Antarctic ice shelves. As opposed to sea ice, these ice shelves are of particular concern because they sit over land, meaning that when they melt, the runoff will end up in the seas, raising sea levels significantly. If the ice sheets in the Antarctic and Greenland disappear entirely, sea level is projected to rise by 65 to 80 meters.[30] Although most scientists agree that this is unlikely to happen before 2100, the AR5 report clearly documents progressive melting: "The sea level equivalent of mass loss from the Greenland and Antarctic ice sheets over the period 1993–2010, has been about 5.9 mm (including 1.7 mm from glaciers around Greenland) and 4.8 mm, respectively."[31] Going forward, the IPCC AR5 report estimates that under its optimistic RCP4.5 scenario (where GHG emissions peak in 2040 and then decline), ice shelves will continue to melt and sea level will rise by an estimated 1.4 meters (range of 0.9 to 2.0) between 2046 and 2065 and 1.8 meters (range of 1.1 to 2.6) between 2081 and 2100. This is clearly not as disturbing as a 65-meter rise in sea levels; nonetheless, these estimates are underpinned by a very conservative model that does not factor in the possibility of runaway warming feedbacks like the process described for permafrost melt.

The third threat—destabilization of the thermohaline circulation—represents a peril that is something out of a science fiction novel. Indeed, the 2004 movie *The Day After Tomorrow* is based on just such a premise—a slowdown and stoppage of the Atlantic meridional overturning circulation (AMOC—part of the global thermohaline circulation).

Just what is this all about? Running silently under all of our major ocean basins is a deep ocean current known as the thermohaline circulation (figure 3.5). This current is powered by thermal and salinity variations and brings heat from warmer equatorial oceans toward the poles and colder water from the poles to equatorial climes. Thanks to this process, nations that are closer to the poles benefit from warmer weather. There is evidence that this process can actually slow or stop, causing sudden cooling in northern climates. Approximately 13,000 years ago, the Canadian provinces of Manitoba and Ontario were dominated by two enormous glacial lakes—Lake Aggasiz and Lake Ojibway. According to one analysis, a retreat of ice in the Hudson Bay region opened a path to the sea for the freshwater contained in these lakes, resulting in a sudden infusion of

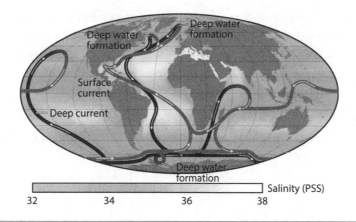

FIGURE 3.5 The thermohaline circulation; PSS = practical salinity scale (Courtesy of NASA—public domain.)

freshwater into the Arctic Ocean. The decrease in salinity in the Arctic Ocean inhibited the Atlantic Ocean's meridional overturning circulation, producing an extended period of climatic cooling in Europe that is known as the Younger Dryas period.[32]

Because of the historical precedent of the Younger Dryas event, concern has been expressed in some circles that a sudden infusion of freshwater from the melting ice caps could trigger a similar event. Before one rushes out to LL Bean to purchase a warmer winter jacket, it should be noted that most scientists dismiss the possibility of this happening (at least in this century, so wait for the sales). According to the IPCC, "it is *very likely* that the Atlantic Meridional Overturning Circulation (AMOC) will weaken over the 21st century." Estimates of weakening range from 11 percent (1 to 24 percent) for the RCP2.6 scenario to 34 percent (12 to 54 percent) for the RCP8.5 scenario.[33] The general consensus appears to be that the rate of ice melt would simply be too slow to catalyze complete collapse. Nevertheless, there is a precedent, and climate science is still too immature to completely dismiss such a fanciful event.

All three of these concerns are not paranoid notions being advanced by people who wear tin foil hats and spend most of their time in root cellars waiting for the end of the world. These concerns are being put forward by scientists of international repute. No mainstream scientist is claiming that these events will transpire in our lifetimes. But as economic activities continue to inject more GHG into the atmosphere, the possibility of a catastrophic systemic crash increases.

Despite the fact that even conservative IPCC projections—let alone those intimating possible catastrophic systemic breakdowns—are dire enough to give us reason to mobilize mitigation efforts, GHG mitigation continues at a snail's pace. Why is there still so much apathy?

UNDERSTANDING THE PSYCHOLOGY OF CLIMATE APATHY

A part of the explanation stems from the confusing way in which impact assessments were made during the halcyon days when people were still trying to decide whether they should be alarmed at the new development that was being called global warming. Another part of the explanation stems from just how difficult it is to put a value on everything that will be impacted by climate change. It is worthwhile to review both factors.

One of the earliest comprehensive economic assessments, and one that is still referred to today, was the *Stern Review on the Economics of Climate Change*, which was released in 2006. In this report, climate change was referred to as "*the greatest and widest-ranging market failure ever seen.*" The *Stern Review* team worked from IPCC climate change impact projections and other scientific assessments to try to economically quantify the projected impacts. The core finding was that climate change would impair global GDP by around 5 percent, every year, now and forever. Moreover, the report warned that if 5°C to 6°C of warming were allowed to occur, this estimate would increase to 20 percent.[34]

The trouble with Stern's reporting strategy was that by putting this estimate into GDP terms, he made it an abstract cost. After all, with global economic growth averaging 2 to 4 percent most years, a 5 percent annual cost does not seem that damning. Yet this is likely not the message that the lead author of the report—Nicholas Stern, the former World Bank chief economist—intended to convey. When one considers that even 5 percent of global GDP in 2007 amounted to approximately US$3 trillion (2010 dollars), one can see that the *Stern Review* was making a far more important point: to avoid US\$3 trillion in economic losses, all that would be required would be a 1 percent investment of GDP in mitigation. In other words, invest US\$600 billion to avoid US\$3 trillion in recurring annual costs.

A second weakness of the *Stern Review* stemmed from the constraints that framed the project. The goal was to quantify economic impacts, not speculate on how these impacts might influence geopolitics and people on the ground. The poorest regions on the earth will experience mass migrations and an amplification of poverty, likely engendering regional strife. Meanwhile, governments in richer nations will be confronted with high adaptation costs; and this financial burden will, in turn, necessitate the imposition of higher levels of taxation or curtailed spending in other areas of governance (healthcare, education, etc.), leading to citizen discontent. In short, estimating the economic cost of climate change involves more than the generation of a number; it is a process that is underpinned by true hardship and geopolitical strife. It is an event that will adversely shape how people in the world interact with each other.

One other caveat (which was acknowledged by the *Stern Review*) was that the study did not estimate health and environmental costs. Yet we will undoubtedly face lives lost, flora and fauna extinctions, and cultural losses. Deaths will rise as people perish in extreme weather events, in the face of droughts, and because of forced migration and flooding as sea levels rise. If climate change is to be fully costed, deaths, extinction, and hardship need to be quantified. Yet economics has no real way to do so without producing contention, which explains why more recent economic estimates have not been able to persuasively supplant the *Stern Review*.

To illustrate just how contentious economic modeling can be, it merits exploring one challenge—estimating loss of human life. There are numerous approaches to this problem, including estimating opportunity cost lost and earning potential foregone. However, one approach that attracts many economists is contingent valuation, which is the process of valuing life by asking how much people would be willing to pay to avoid dying (or prevent a loved one from dying) because of climate change impacts. One way is to begin with a question such as "If you had the power to do so, how much would you be willing to pay to give your loved one ten extra years of life?"

Economists then aggregate the responses and calculate an average to ascertain the value of averting premature deaths. This value can then be multiplied by the estimated number of premature deaths attributed to climate change to come up with a total cost estimate. As morally repugnant as this method may be, the results of such an exercise demonstrate just how contentious climate change impact projections can be if factors such as lives lost, flora and fauna extinctions, and health costs are factored into the analysis.

Let's imagine that through contingent valuation, the average value of a human life is estimated at US$1 million. How many deaths can we expect? A recent study published in *The Lancet* estimates that impairment of food production will result in 529,000 premature deaths per year between 2010 and 2050.[35] The World Health Organization estimates that climate change–induced deaths from malnutrition, malaria, diarrhea, and heat stress will amount to approximately 250,000 per year by 2030.[36] Given just these two estimates and recognizing that deaths from extreme weather events have yet to be factored into the equation, it would not be an exaggeration to conclude that climate change can be expected to cause over 1 million premature deaths per year over the next seventy-five years. Using our metric of $1 million per life, the price tag for these deaths would amount to US$1 trillion annually. In other words, premature deaths alone would inflate the *Stern Review* estimates by 33 percent.

Flora and fauna extinctions will also amplify as climate change intensifies. In a study published in *Science*, it was estimated that extinctions stemming from the RCP8.5 trajectory (IPCC's business-as-usual scenario) could amount to as many as one in six species by 2100. Many of us see value in the existence of another species that goes beyond economic value. So again, employing contingent valuation theory, how much would you be willing to pay to save these species? Let's assume that after undertaking this exercise, it turns out that humans are willing to pay US$100 per person per year to ensure that species do not go extinct. In aggregate, with 7.6 billion people on the planet, this total would come to US$760 billion per year, adding a further 25 percent to the *Stern Review* estimates.

Finally, there will be cultural losses as entire villages are destroyed by extreme weather events or rendered inhabitable because of environmental change (flooding, degraded soil, etc.). Once again using contingent valuation, how much would you pay to have your birth home preserved, or to save the cemetery in which your ancestors are buried? Costs such as these would need to be added to Stern's financial estimates to get a true picture of the economic damage associated with climate change. After investigating just three of the omitted variables, it should be apparent that the *Stern Review*'s estimates are extremely conservative. Adding all unaccounted-for costs could easily double the bill.

Contention over how to properly value climate change impacts is manna from heaven for vested interests. All of this uncertainty provides a window of opportunity to sow seeds of doubt and denial for those who seek to maintain the status quo. But what type of person would want to preserve the status quo amid projections of global economic, ecological, and social disaster? We consider this question in the next section.

THE MUDDYING EFFECT OF VESTED INTERESTS

As mentioned in chapter 2, energy expert Travis Bradford estimates that the economic activity generated by the energy sector amounts to 7 to 10 percent of global GDP.[37] Put another way, as much as $1 in every $10 to $14 that is in your pocket now will eventually make it to some energy provider somewhere. In laying out his vision for electric mobility, the entrepreneur Shai Agassi came to a similar conclusion in a 2007 white paper:

> The total economic dislocation [by electric mobility] seems almost incomprehensible. Fuel at the pump represents a market of $1.5 trillion every year. Cars and components size roughly to the same size of market, $1.5 trillion a year. Financing for new cars, gaining acceptance worldwide is estimated at $0.5 trillion a year. Clean electricity generation for cars is a market that will reach $0.15 trillion a year. ERG infrastructure construction will reach levels of $0.5 trillion a year. Battery manufacturing will reach similar levels of $0.5 trillion a year, accounting for reduction in battery cost as the market size will continue to increase. In-car services, such as GPS, media, phone as well as related services such as insurance and maintenance collectively worth more than $1.5 trillion a year will be

affected. Carbon credits alone will be worth roughly $0.3 trillion when all cars are driven on clean electricity. In the aggregate, we are looking at an annual dislocation reaching roughly $6 trillion a year.[38]

Most people find these figures hard to accept because the money they spend on household heating, electricity, and gas for the cars does not amount to 10 percent (or more) of their monthly expenditures. The discrepancy is explained by the notion of embedded energy. Every product and service that we consume requires energy. The last time you went to a restaurant, you likely sat in a chair that required energy to produce, ordered from a menu that required energy to create, benefited from electricity that lit the lights and allowed you to read from your menu, enjoyed a meal that required energy to cook, and consumed food that required energy to produce. All of these energy expenditures were incorporated into the cost of the meal that you enjoyed—it just wasn't obvious. Indeed, energy lurks in the background of everything that human beings do. It is ubiquitous but often unnoticed.

Some very powerful firms and wealthy individuals have benefited from profits in the energy sector. Consider the financial might of the world's largest energy firm—ExxonMobil Corporation (Exxon). If one were to compare the 2016 revenues of Exxon—US$197.5 billion[39]—to the GDP of nations around the world, only 47 nations would have a higher GDP than Exxon's 2016 reported revenue.[40] With revenue in 2016 of US$173.1 billion,[41] Royal Dutch Shell generated more revenue than 51 nations, falling just behind New Zealand with a GDP in 2016 of US$185 billion. In the face of climate change mitigation efforts, these firms face a real and present threat to their continued existence at the upper levels of the economic oil pyramid.

The transport sector also boasts firms that possess a significant amount of financial might. Petrol companies such as Chevron, BP, Shell, and Amoco have established franchise petrol stations around the world. BP alone boasts over 74,000 employees, approximately 18,000 service stations, and operates in seventy-two nations.[42] However, it is not just the petrol providers who face investment risk. Automobile manufacturers are some of the most entrenched industrial entities in the world because of the cost of plants and service support networks. The largest automobile manufacturer—Toyota—posted 2016 revenues of US$236.7 billion, making it larger than

Exxon based on revenues.[43] Fortunately for Toyota, its market leadership is predicated on leadership in hybrid and electric vehicle development. Not all automobile manufacturers are so forward thinking, and so there are fortunes to lose amid a transportation industry transition.

In addition, there are powerful trucking firms (such as Daimler AG, which manufactures Mercedes-Benz and Freightliner trucks), heavy equipment manufacturers (such as Komatsu and Caterpillar), auto parts manufacturers (such as Valvoline, Pennzoil, and Bohai Piston), motorcycle manufacturers (such as Honda, Yamaha, and Kawasaki), and even manufacturers of boat motors (such as Evinrude and Mercury), all of which stand to lose revenues if internal combustion engines were to be replaced by electric vehicles.

Even if these firms were unilaterally motivated to alter their respective business strategies to extract themselves from product lines that depend on fossil fuel technologies, a phaseout would take considerable time. All of these manufacturing firms have factories that represent sunken investment. These factories are for the most part custom-designed facilities that are not easily re-tasked for producing other products. The extent of entrenched investment can be somewhat approximated by a review of the balance sheets of many of these firms. Exxon, for example, reported long-term investments and fixed assets amounting to US$279.3 billion in 2016.[44] Long-term investments and fixed assets in 2016 for Ford, Daimler AG, Caterpillar, and Honda amounted to US$114 billion, US$180 billion, US$48 billion, and US$89 billion, respectively.[45]

What this should suggest is that for many firms that have a financial stake in businesses connected to fossil fuel products, there are financial incentives to delay any transition to a new business model. Sunken investment first needs to be written down, and these firms need to be sure that their alternative business models will allow them to retain market leadership. Despite current market leadership, survival is not guaranteed amid market transformation. The graveyard of bankrupt and floundering businesses is littered with the names of powerful firms that could not successfully make such a transition—Kodak film, Brother typewriter, Commodore computers, Nokia, and Blackberry, to name but a few.

Perhaps understandably, then, many of these firms do what they can to try to delay change. Some firms choose to do so through the political process by employing lobbyists to manipulate political will. For example, looking at funding for the top 40 lobby groups in the United States in 2016, Southern Company (electricity and natural gas) ranked #13 at US$13.9 million, Exxon was #19 at US$11.8 million, Koch Industries was #26 at US$9.8 million, Royal Dutch Shell ranked #33 at US$9 million, and General Motors was #39 at US$8.5 million.[46] Other firms choose to respond strategically by attempting to diversify business streams. For example, Royal Dutch Shell's New Energies division invested just under US$200 million in 2016 but plans to ramp up investment to US$1 billion by 2020.[47] Still other firms seek to prevent change by minimizing the adverse impact of their products on climate change. Most automobile manufacturers have divisions that are attempting to create either hybrid vehicles or electric vehicles, but some are more proactive than others. For example, in 2016, Volkswagen sold over 60,000 electric and hybrid-electric vehicles. In contrast, Ford sold approximately 25,000 such vehicles.[48] In the same year, Toyota sold 1.4 million hybrids.

In summary, entrenched industry resistance can be best understood as an incentivization problem. When the renewable energy market was emerging, traditional market leaders were more active in resisting competitive advances and less active in undertaking research and development to join the transition. As the market continues to evolve, the strategy can be expected to reverse, with traditional market leaders becoming more active in undertaking research and development and less inclined to resist technological change. For example, in 2018, electric vehicle (EV) market laggard Volkswagen announced that by 2022, it would have sixteen locations around the world for producing EVs and was aiming to produce 3 million EVs annually by 2025.[49] However, this does not mean that opposition will disappear. As the next section documents, in this high-stakes environment, some stakeholders are prepared to stretch the bounds of ethical practice.

PATH DEPENDENCE AND RISING EMISSIONS

In 2008, James Hansen, then director of the NASA Goddard Institute for Space Studies, penned an article to mark the twentieth anniversary of his testimony before the U.S. Congress, which was widely considered to have put climate change on the political map in the United States.[50] In this article, he stated, "CEOs of fossil energy companies know what they are doing and are aware of long-term consequences of continued business as usual. In my opinion, these CEOs should be tried for high crimes against humanity and nature."

Although this is a shockingly blunt response from a respected scientist, one can understand his frustration. Efforts to delay a transition by manipulating public support have been so successful that, up until very recently, many feared that we would not see a shift to alternative energy forms in this generation. Increasingly the impact of atmospheric warming will be felt—change will be unavoidable. Yet GHG emissions, which continue to be emitted as fossil fuel interests strive to extend their profits, will remain in our atmosphere for decades—even centuries for some substances.

There is clear evidence that fossil fuel interests have funded misinformation campaigns. In an investigation entitled *The Climate Deception Dossiers* published by the Union of Concerned Scientists, researchers released seven "deception dossiers"—collections containing some eighty-five internal company and trade association documents that have been "leaked to the public, come to light through lawsuits, or been disclosed through Freedom of Information Act (FOIA) requests," which provide evidence of "a coordinated campaign underwritten by the world's major fossil fuel companies and their allies to spread climate misinformation and block climate action."[51] According to a report from *MediaMatters*, on the heels of a January 2016 NASA report that 2015 was the hottest year on record, "big oil ads outpaced climate-related coverage on CNN by nearly 5-to-1."[52] It was impressive strategizing on the part of big oil, but not so impressive for citizens who are concerned about the perils of climate change.

With that said, only part of the blame rests on the shoulders of firms connected to the fossil fuel industry. After all, they are acting in a manner that one would intuitively expect of a sunset industry. A far more insidious enabler of all this is evidence of widespread dumbing down of society. As author Mark Baer described the world leader in this category, "There is a growing and disturbing trend of anti-intellectual elitism in American culture. It's the dismissal of science, the arts, and humanities and their replacement by entertainment, self-righteousness, ignorance, and deliberate gullibility."[53] There is anecdotal evidence that the anti-intellectual spirit that has gripped the United States is also alive and kicking in Europe.[54]

In all fairness, another enabler of misinformation relates to information overload. We are in an information age, with factoids at our fingertips. Just say "Okay Google" or "Hey Siri" and your questions can be answered. Thanks to the reach of the Internet, there has been an explosion of media outlets, political bloggers, and social media sites where misinformation and inaccuracies reign. Any yahoo with an Internet connection can now become a political pundit, commenting on what he or she perceives to be the way forward for the energy sector. These wayward comments are then picked up by other sites and reported as truths, further inhibiting the ability of people to suitably scrutinize statements. With the growing anti-intellectual spirit and so many demands on our time, people are not taking the time to scrutinize trends or understand causes.

Busy citizens now receive bits and bytes in a steady stream. Most of it goes in one ear and out the other unless the story has been sensationalized. So reports that lay claim to conspiracies or that challenge existing beliefs are popularized. In many nations, consumers have become vacuous sponges of sensationalist hyperbole and politically motivated propaganda.

Meanwhile, to the glee of fossil fuel interests, the escalation of demand for energy in many developing nations has spurred an unprecedented wave of expansion in electricity generation facilities, the majority of which are still coal-fired power plants. Supply-side expansion of coal-generating facilities and increased demand for automobile ownership have more than negated international efforts to reduce GHG emissions. For example, compared to 1990 levels, the European Union (EU) has managed to decrease annual GHG emissions by 24 percent to 4,282 million tonnes of CO_2 equivalent $(MtCO_2e)$.[55] Unfortunately, this 1,383 Mt reduction has been trivialized by China alone, which saw annual GHG emissions increased by a whopping 3,387 Mt between 1990 and 2013.[56] Despite the perils of impending climate change disaster, we are still seeing signs of global regression in GHG emissions—one step forward, two steps back.

In addition to the catalytic effect of economic development in driving GHG emission increases, population growth promises to exert further pressures on climate change mitigation efforts. As of November 2017, the world's population stood at 7.6 billion people.[57] By 2030, the United Nations predicts that the world's population will have increased by 12 percent to 8.5 billion.[58] In other words, if projections are correct, in thirteen years we will have added another Indonesia, Brazil, Pakistan, Nigeria, and Bangladesh to the global mix of people who are competing for scarce resources.

The pairing of amplified economic growth and burgeoning population numbers hits on two fronts when it comes to climate change mitigation. First, as just discussed, these two forces are driving demand for energy supplies to levels where no one technology can keep up. Prices for all fossil fuels have been on the rise as a result of demand-side pressures.[59] Second, these factors are also driving deforestation. Globally, forests play an invaluable role as carbon sinks. It has been estimated that forests sequester up to 30 percent of all anthropogenic GHG emissions.[60] When trees are cut down, not only do the sinks disappear, but the CO_2 that has been absorbed by the trees is then released to the atmosphere.

In the face of amplified population growth and an economic paradigm that fails to distinguish between benign and malignant economic growth, there are valid concerns that we are in the process of rendering a nearly intractable problem totally intractable. Clearly there is a degree of exigence needed to avert the worst perils attributed to climate change. Therefore, efforts of the international community in securing consensus and commitment to aggressive GHG reduction efforts merit review. The latest benchmark in this regard is the Paris Agreement.

WE'LL ALWAYS HAVE PARIS

Even before the UN Framework Convention on Climate Change (UNFCCC) was adopted in 1992, there was widespread disagreement over obligations that each nation should have in mitigating GHG emissions. For twenty-five years, international negotiations surrounding climate change mitigation have lurched forward like a drunken sailor who is not sure whether he should stumble home or return to the pub from whence he came. The initiatives, fostered through the auspices of the Kyoto Protocol, seeded the development of a number of renewable energy policies and investments around the world; however, in aggregate, these successes have had limited impact in the face of progressive escalation of energy demand. Fossil fuels still rule the roost.

As negotiations proceeded on setting targets for the second commitment period (2012 to 2016) under the Kyoto Protocol, it appeared as if the international community was losing the stomach to reconcile the ideological differences exhibited by the member states. The annual Conference of Parties (COP) meetings that were intended to solidify action came and went with very little concrete progress toward binding emission reduction commitments by all parties. At the 2009 COP conference in Copenhagen, there was hope of concrete targets being set; but at the last minute, ideological rifts reemerged and the only achievement was the signing of a political promise to constrain carbon emissions to mitigate climate change.[61] Subsequent to Copenhagen, meetings in Mexico, South Africa, Qatar, Poland, and Peru provided some scenic backdrops for photo opportunities to serve as testament that something was being discussed, but, once again, promise of progress was evanescent at the best. Then along came the 2015 COP meeting in Paris and adoption of the Paris Agreement.

The Paris Agreement was part of the UNFCCC process whereby ratifying nations agreed to commit to the establishment and achievement of "nationally determined contributions" (NDCs) to GHG abatement. While there are no mandates as to what these targets should be set at, the agreement stipulates that the parties should together seek aggregate reductions in GHG emissions to keep global warming well under 2°C and that each subsequent round of NDCs should seek to achieve progressively greater GHG reductions. As of December 21, 2018, 184 of the 197 parties to the UNFCCC convention ratified the Paris Agreement.[62]

Although it is conceivable that nations could establish artificially high NDCs in order to be seen as a member in good standing without actually having to make an effort, perhaps the most surprising development is how assertive many of the nations have been in setting emission reduction targets. For example, in 2017 France announced plans to phase out the use of coal in all power plants by 2022 and to ban the use of petrol and diesel vehicles by 2040.[63] The EU, as a bloc, has agreed to a 2030 binding target to cut emissions in the EU by at least 40 percent below 1990 levels.[64]

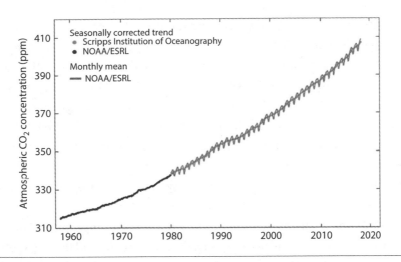

FIGURE 3.6 Trends in carbon dioxide emissions, 1960–2018 (1980–2018 monthly data from the National Oceanic and Atmospheric Administration, Earth System Research Laboratory [NOAA/ESRL] [Dlugokencky and Tans, 2018], based on an average of direct atmospheric CO_2 measurements from multiple stations in the marine boundary layer [Masarie and Tans, 1995]; 1958–1979 monthly data from the Scripps Institution of Oceanography [Keeling et al., 1976].)

On the other hand, the Paris Agreement has been criticized as being yet another toothless international agreement. Former NASA scientist James Hansen calls the process "all talk and no action." He argues that it lacks a binding enforcement mechanism and that its success is entirely dependent on the individual efforts of member states.[65] Looking purely at GHG emission trends (figure 3.6), he seems to have a point. Global CO_2 emissions from fossil fuels in 2018 are forecast to be 2.7 percent greater than in 2017, which was also a year of significant growth.[66]

In June 2017, U.S. president Donald Trump announced that the United States would withdraw from the Paris Agreement at the earliest possible time, which under article 28 of the agreement means that the United States would have to remain a party to the agreement until November 4, 2020. Nevertheless, with the United States having announced its intentions, there are concerns that this will give rise to an erosion of support for the Paris Agreement.

Although the question of whether the Paris Agreement will continue to engender strong commitment from ratifying parties is of some importance, we would contend that the same forces that led to the ratification of the Paris Agreement by 184 nations will drive progressive emission reductions, regardless of the continued viability of this agreement. The basic premise is that 184 nations have agreed to commit to progressively strong NDCs for the economic, environmental, and sociopolitical reasons explicated earlier. The ten energy transition drivers outlined in chapter 1 will foster a shift to decarbonized energy systems regardless of whether international agreements exist.

Moreover, as the predictive capability of climate change impact models congeals and climate change events rise in their ferocity and impact, special interests will have a harder time influencing political will to impede a transition. In 2017 alone, six major hurricanes hit the United States. Joel Myers, founder of AccuWeather, estimates that the damage from the two largest of these (Hurricanes Harvey and Irma) will amount to US$290 billion, or 1.5 percent of U.S. GDP—effectively nullifying any economic gains for the period.[67] These sensational disasters have forced media outlets to interrupt reruns of *Duck Dynasty* and the exploits of *Honey Boo Boo* and have forced even the most apathetic citizens to ponder why this is all happening.

CONCLUSION

In conclusion, then, even though we can expect continued efforts, often disingenuously, by fossil fuel interests to endeavor to preserve the status quo, their successes at the turn of the century will likely not be replicated going forward. Scientific understanding has improved, and people are starting to see through their own eyes that climate change skeptics may be both biased and poorly informed.

Although many energy firms are privately held and this situation allows them to develop strategy in any way that they deem fit, it is also worth noting that a number of the largest energy firms are publicly traded. This means that their corporate strategies are vetted by boards made up of investors. These investors come from the same pool of citizenry that suffer in the face of extreme weather events and see the impact of climate change unfolding. Consequently, corporate boards are increasingly compelling senior management to make more sustainable business decisions. A prominent case in point was a 2016 mandate supported by 38 percent of Exxon shareholders in two prominent U.S. financial portfolios that the firm must annually assess the portfolio impacts of climate change policies.[68] A mere year later, at the Exxon annual shareholders meeting in 2017, investors controlling 62 percent of Exxon shares voted in favor of this measure.[69]

In a nutshell, we are witnessing a potential paradigm shift, which is broadly defined as the adoption of a worldview that differs from and supplants a worldview that primarily guided human endeavor. As Thomas Kuhn explains in his analysis of how paradigm shifts occur, resistance from the parties who benefit from the status quo continues to inhibit change until scientific evidence is sufficient enough to defang opposition to change.[70] This is not a speedy process.

One of the more famous examples of a paradigm shift illustrates this point. Shortly before his death in 1543, Copernicus published a book entitled *On the Revolutions of the Celestial Spheres*, where he argued that the accepted paradigm of the earth being at the center of the known universe was incorrect and it was the sun rather than the earth that was at the center.[71] Opposition to this perspective was led by the Catholic Church, which was worried that this perspective would undermine religious doctrine concerning the origin of our planet. Attempts to refute or disparage this theory were carried out for centuries. Almost a hundred years after the death of Copernicus, the scientist Galileo Galilei was imprisoned in 1633 for "vehement suspicion of heresy" in supporting Copernican views.[72] It wasn't until 1835 that Copernicus's book was dropped from the church's index of prohibited books.[73]

The point is, when it comes to climate change, science will continue to be refined and climate change projections will become more precise. Meanwhile, extreme weather events and other climate change–related damage will continue to escalate. Members of the general public will begin to see for themselves what the scientific community has been attempting to communicate for the past forty years: climate change is real and a major threat. As all this happens, vested interests will have a harder time influencing public opinion and swaying political support. While this change in mindset is occurring, the cost of renewable energy technology will continue to fall, rendering the final economic arguments in favor of the status quo moot. Yet this will not happen tomorrow. The struggle will continue, and the clock is ticking on climate change.

To expedite this transition, each person who has been capable of understanding the science underpinning climate change projections needs to take more responsibility for invoking change. There are innumerable activities that the average citizen can undertake to ensure that this paradigm shift does not take three hundred years to occur. Here are just a few:

- Write to your political representatives to let them know that you are supportive of climate change mitigation policies.
- Form community groups to help disseminate knowledge on climate change and encourage the adoption of lifestyle choices that will not exacerbate GHG emissions.
- During elections, ensure that you vote for politicians who support climate change mitigation and, if possible, make the effort to lend support to the campaigns of such politicians.
- Embrace small initiatives such as purchasing carbon offsets, supporting tree-planting programs, or encouraging energy efficiency. Moreover, enlist the participation of others.
- Help to organize cooperative investment efforts in GHG mitigation when government efforts are failing or nonexistent.
- Hold the media accountable by lodging complaints or writing letters to the editors when encountering irresponsible reporting.
- Develop an online presence to counter misinformation campaigns or to disseminate information to others.

It is frustrating when change is only achievable through daily endeavors that receive little reward and provide limited evidence that a difference is being made. However, in many ways, playing a role in helping to reduce GHG emissions might wind up being the most significant accomplishment that we can lay claim to in our careers. Consider how much can be accomplished through grassroots efforts. If 1,000 of us set out to positively mobilize 100 people a year to act in support of climate change mitigation, over 10 years, we would have mobilized 1 million people. If those 1 million did the same thing, in 10 more years, 1 billion people would be mobilized to act in support of climate change mitigation. In other words, in 20 years, a small group of grassroots advocates could engender substantive change. Fortunately for us, it is not a difficult sales job. Most people already know that the earth revolves around the sun and that our activities are causing that sun to warm our planet. Now all we need to do is to show them why we might want to ensure that others know about this as well.

Even the Catholic Church, which shunned Copernicus almost five hundred years ago, has weighed in on this paradigm shift. Pope Francis in a statement in 2017 had this to say on the issue: "Man is a fool, a stubborn man who will not see. . . . Anyone who denies [climate change] should go to the scientists and ask them. They speak very clearly . . . climate change is having an effect, and scientists are telling us which path to follow. And we have a responsibility—all of us. Everyone, great or small, has a moral responsibility . . . we must take it seriously . . . history will judge our decision."[74]

Fortunately for humanity, taking an active role in this change is not necessarily a huge sacrifice for one simple reason: the cost of alternative technologies has dropped precipitously, rendering the decision to adopt renewable technologies an economically sensible process. Moreover, the commercial opportunities that are emerging as the transition takes hold favor decentralized, highly productive enterprises. This transition comes with a silver lining in that the "Great Energy Transition" might be exactly what is needed to help redirect commerce from a model of big business that hides inefficiencies through economies of scale and resource exploitation to a model of small and medium enterprises that extract as much value and productivity as possible from each resource employed. As Tony Bennett once crooned, "The best is yet to come." We will see why this is the case in the next chapter.

4

Managing Uncertainties While Promoting Technological Evolution

I t is easy to find oneself caught in the middle, frustrated both by those whose vision is limited by the present mindset and by those who rush to a future fantasy without understanding what it might cost, the time it could take, and the transitionary losses incurred in getting there. On the one hand, we appreciate that visionary thinking can distort expectations and motivate premature actions; on the other hand, myopic strategies can miss or at least fog windows of opportunity. Roy Amara, past president of the Institute for the Future, summed up the problem aptly: "We tend to overestimate the effect of a technology in the short run and underestimate the effect in the long run."[1] Thus, while this chapter is full of facts, figures, and assessments of the present state of technology along with current trends, hopefully readers will also gain a sense of the overarching uncertainty that permeates the challenge of planning for the future. Even though the cost of renewable energy technologies is dropping precipitously and battery storage technologies are progressing in leaps and bounds, the transition is fraught with uncertainties that must be handled with kid gloves.

This chapter explores strategies that governments could use to govern technological evolution and the profound uncertainties that surround possible energy transition paths. In keeping with our desired conceptualization of energy planning as a type of community planning, we emphasize the need to capitalize on the smart energy advances that are transforming how we view electricity generation and to map the role of homes, businesses, and mobility systems in this transformation. A plethora of recent smart energy products and approaches has been newly enabled by increased connectivity, digitalization, and decentralization. Furthermore, they have been made accessible by novel forms of collaborative adoption. Thus, we begin by examining these technology trends and possible energy futures. Then we consider the strategies available to renew and align energy infrastructure to support smart grids, homes, businesses, and mobility systems. The chapter ends by exploring how collaborative adoption could make these new energy infrastructures widely and swiftly available across heterogeneous segments of society.

Electrification, digitalization, and decentralization are transforming energy systems worldwide.[2] Electrification is the great enabler of an advanced energy economy. In much of the world, electricity is supplanting portable fuel (oil, natural gas, and coal) as the energy conduit for a wide array of services, including light, comfort, mobility, and heat, and this trend is expected to accelerate.[3] Digitalization provides real-time automated communication to optimize the operation of systems using network technologies, the Internet of Things (IoT),[4] remote controls, automation, and a heavy dose of smart sensors and meters. Decentralization is on the rise, buoyed by advances in the performance of systems that customers can actively operate, including rooftop solar photovoltaic (PV) panels, energy efficiency, demand response, electric vehicles, and energy storage. These three technology trends are causing drastic changes in national economies and societies, with knock-on effects that are reverberating across the business landscape and impacting the daily lives of individuals. The future is not what it used to be.

In combination, electrification, digitalization, and decentralization have enabled the rapid onset of the information age, the fourth industrial revolution, and collaborative consumption. New social and business models that reconceptualize energy services are leveraging these new capabilities—optimizing aggregation platforms, smart devices, and the IoT. These models are diffusing rapidly.

In the wake of these transformations, tension is building between supporters of the status quo and advocates of alternative futures. This tension is inevitable because conventional energy technologies do not seem compatible with this new paradigm. Coal and nuclear power tend to be clunky technologies compared to solar power applications, storage of energy, and distributed control over energy systems. The traditional ownership of homes, cars, and electricity-consuming assets is being challenged by new business models that offer alternative means to satisfy everyday needs and demands that don't involve stenciling your name on purchased products. Digitalization platforms supported by information and communication networks are enabling a shift toward open markets and shared ownership of distributed assets. For example, modern economies across the world are experiencing (1) the distributed provision of electricity (e.g., leased solar rooftops and community solar PV), (2) bike-, ride-, and car-sharing services (e.g., Uber, Lyft), (3) on-demand rental units in homes (e.g., Airbnb), and (4) distributed water services (e.g., rainwater collection, management and storage integrated into buildings). Markets are being created for sharing all kinds of gadgets—even for swapping clothes and recycling food.

What do such trends mean for the great energy transition, and what strategies can governments use to renew and align energy infrastructure to support such trends? Or is caution in order, to ensure that the outcomes are indeed desirable and sustainable? That is the focus of this chapter. After presenting the case for expanded electrification, we turn to a discussion of digitalization. We then describe the decentralization of energy resources and their integration into smart systems that deploy collaborative consumption as an underlying business premise.

ELECTRIFICATION

Electricity is becoming the dominant energy carrier across the globe (figure 4.1).[5] In industrialized nations, it is the dominant source of energy for meeting the service requirements of homes and offices: providing requisite

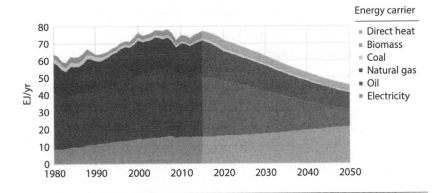

FIGURE 4.1 The electrification trend in North America: final energy demand by carrier (DNV-GL, *Energy Transition Outlook 2017*, https://eto.dnvgl.com/2017 /main-report.)

power for air conditioning, lighting, computers, communications, and an increasing array of electrical appliances, gadgets, and accessories. Even in advanced nations, demand for electricity is expected to expand rapidly with the growth of data centers that provide large-scale cloud services, advanced electric space and water heaters, changes in industrial processes (with the growth of electric arc furnaces and additive manufacturing),[6] and enhanced electric vehicle usage (including electric cars, delivery trucks, bikes, and mopeds). The reach of electricity grids and per capita rates of consumption will also expand considerably as large-scale electrification efforts continue around the world, aiming to bring light and power to the more than 1 billion people who currently lack it.[7]

Technological innovations tend to be at the center of enhanced reliance on electrification. Heat pumps are one of the most promising technologies for meeting the growing demand for space heating and cooling.[8] For decades, traditional heat pumps have been used to heat and cool buildings, and more recently, heat pump water heaters have also gained market share. To this mix comes a new generation of super-efficient integrated air-source heat pump systems that can heat and cool spaces while also heating water and dehumidifying indoor air. These innovative heat pump systems will economically justify conversions from fossil fuel–driven heating systems in commercial buildings.[9] Examples include the Rebel heat pump introduced by Daikin Industries in 2012 and the Weather Expert system introduced by Carrier/United Technologies in 2013; both offer previously unmatched levels of energy efficiency. Until recently, air-source heat pumps were not able to meet the needs of homes in cold climates. But with the development of an improved scroll compressor and a tandem compressor design, air-source heat pumps can now meet the heating needs of homes in the coldest regions of the United States.[10] With these new electric technologies, building owners can save money and reduce their energy consumption while reducing carbon footprints.[11]

Globally, between 1980 and 2010, the energy used for heating and cooling grew by an estimated 39 percent and 61 percent, respectively, in homes and businesses, and the demand for space cooling is expected to nearly double by midcentury. In residential buildings, the growing number of households and the increasing size of homes are contributing to this growth.[12] In commercial buildings, a twofold increase of global GDP is expected to expand the commercial building stock and its energy requirements. As of 2018, heating and cooling systems consume nearly half of global total final energy consumption.[13] Therefore, the impact of a transition to new electric space-heating technologies in terms of greenhouse gas (GHG) emission offsets could be sizable.

While energy efficiency and technological improvements will likely reduce the energy required per square meter of building space in the developed world, these advances may not be sufficient to offset the burgeoning global demand for cooling.[14] Of particular concern is the need to bring electricity (and other modern energy services) to the hundreds of millions of unconnected homes and the impact that such expansion will have on the demand for fossil fuels.

Over the long run, climate change is expected to significantly increase the demand for space cooling while slightly decreasing the demand for space heating.[15] Because electricity is the dominant power source for supporting air conditioning, a warming climate is anticipated to increase the demand for electricity. Rising temperatures could push the electric grid to the brink in some countries. For example, in the United States, warming trends could dictate an investment of $180 billion worth of new power plants this century to meet the increasing need for cooling.[16] A 1.2°C increase in temperature in the United States could cause primary energy use to increase by 2 percent in 2025.[17] Globally, much larger energy demand impacts due to air conditioning technologies have been projected by the middle and end of this century.[18] Consequently, relative to a future with no climate change, larger increases in peak demand for electricity are anticipated, necessitating a 14 to 23 percent increase in electricity capacity additions between 2010 and 2055[19] and a further 10 to 25 percent increase by the end of the century.[20] Such developments exert additional stress on the electric grid, which is already vulnerable to climate-related outages.

These changes are expected in developed nations that already enjoy a strong electricity infrastructure. In developing nations, the transitional requirements will be far greater because many citizens in the developing world are not yet even connected to electricity grids. As of 2017, there were still 1.1 billion people without access to electricity[21] and 2.8 billion without access to clean cooking facilities.[22] As more households are connected predominantly through the same technologies already described, we can expect an avalanche of demand for electricity infrastructure and electricity power sources.

As heat pumps, electric cars, additive manufacturing, and other such drivers of amplified electrification continue to take hold across the world, the climate impact could be perilous if electric power comes from fossil fuels. Conversely, benefits could be substantial if the transition is supported by a low-carbon electric power mix. Under such a transition, electrification will support further digitalization, potentially giving rise to a new facet of the information age.

DIGITIZATION AND THE INFORMATION AGE

The increasingly digitalized economy has the potential to alter the workings of all major infrastructure components. In the IoT era, sensors and actuators can be embedded into devices, vehicles, homes, factories, and appliances, connecting all six forms of capital—the social, human, financial, institutional, physical, and natural—in a distributed urban network. Enhanced digitalization enables "prosumers" of services to interact in a new sharing economy, tearing down the walls of the economic paradigm that tradition-ally separates producers from consumers. In this new paradigm, consumers become producers too. From the top down, public agencies can use the IoT and big data analytics to plan infrastructure that amplifies responsiveness and efficiency, and from the bottom up, prosumers can use information and communication technologies to foster democratic governance and shape infrastructure investments that better reflect their preferences.

One of the key digitalization technologies that is transforming the electric grid is the smart meter. Metering technologies have evolved from electro-mechanical meters that must be physically read on site, to one-way auto-mated meters that transmit data digitally back to the electricity provider, to an even more sophisticated form—two-way automated meters that transmit consumption information back to the consumer as well as back to the elec-tricity provider.[23] Smart meters can measure and record energy usage data at hourly intervals or more frequently, and they can provide usage data to both consumers and energy companies on a real-time basis. They enable us to understand usage trends, pinpoint areas of inefficiency, and even sell off excess energy that is produced in homes, businesses, and industrial plants.

DISTRIBUTED ENERGY RESOURCES (DER)

The electric power industry is experiencing a significant transformation in the way electricity is generated and delivered throughout the grid. Distrib-uted energy resources (DER) (also sometimes referred to as "decentralization of energy resources") are attracting a growing amount of capital investment and will become more important as consumers and states increasingly value choice, resilience, and clean energy resources.[24] For example, consider just these trends in the United States:

- More than 14 million electricity customers are supplying power back into the grid.[25]
- More than 80 GW of combined heat and power generation facilities, accounting for more than 8 percent of total U.S. generating capacity, are operated by commercial and industrial customers.[26]

- Distributed solar capacity nearly doubled from 7.3 GW in 2014 to 13.8 GW in February 2017.[27]
- More than 16 million customers in the United States participate in wholesale or utility demand response or time-varying rate programs.[28]
- Millions more consumers maintain backup generators or have installed energy storage systems.
- The charging cycles of roughly 542,000 electric vehicles are being managed, providing yet another distributed energy resource.[29]

Additionally, a growing number of utilities are relying on distributed generation or storage to avoid more costly grid investments.[30] Some are also relying on distributed power electronics, operating on a subcycle basis, to optimize voltage and reduce generation requirements. This is a period of significant power system innovation and technology transformation in which the underlying trend is toward a flattening out of energy infrastructure. No longer is power generated inside a large grey edifice atop a hill and delivered down to the masses. Increasingly, the masses are becoming power plants themselves.

Globally, the promise of DER is heartening, given the amplified demand for modern energy services. DER systems can serve as a complement to centralized energy generation systems, or as a substitute. In remote developing communities, they can facilitate affordable lighting, enhance communications, and improve the quality and availability of education by supporting evening studies and the enhanced use of informational technologies in the classroom. DER can also enable greater quality and availability of health services, bringing light, power, and interconnectivity to medical clinics, for instance. DER systems, as well as the hybridization of existing mini-grids, may also reduce dependence on fossil fuel imports. They offer an unprecedented opportunity to accelerate the transition to modern energy services in remote and rural areas, while also offering the following co-benefits:[31]

- Cost savings when compared to grid services in many markets[32]
- Enhanced fuel availability and/or stability and predictability of prices[33]

- Modularity, flexibility, and rapid construction times[34]
- Faster technological learning curves and rates of improvement compared to fossil fuels[35]
- Enhanced reliability and resilience[36]
- Improved health through reductions in indoor air pollution[37]
- A contribution to climate change mitigation[38]
- Attenuation of deforestation and degradation of the environment[39]
- Positive effects on female empowerment[40]
- Reductions of poverty among vulnerable groups[41]

In recent years, the off-grid solar power sector in particular has been one of the fastest-growing sectors.[42] By 2016, more than one hundred companies worldwide actively focused on stand-alone solar lanterns and solar home system kits.[43] Since 2013, at least 125 million off-grid solar systems have been sold worldwide. As figure 4.2 outlines, in 2017 about 26 million off-grid solar systems were sold.[44]

The deployment of renewable mini-grids also accelerated in 2016, expanding to a global market now worth more than US$200 billion annually,[45] with installations on almost every continent.[46] Mini-grid projects are being implemented with increasing emphasis on interconnection: integration

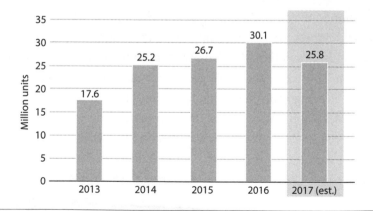

FIGURE 4.2 Annual global sales of off-grid solar systems, 2013–2017 (Renewable Energy Policy Network for the 21st Century [REN21], *Renewables 2018 Global Status Report*, http://www.ren21.net/gsr-2018/.)

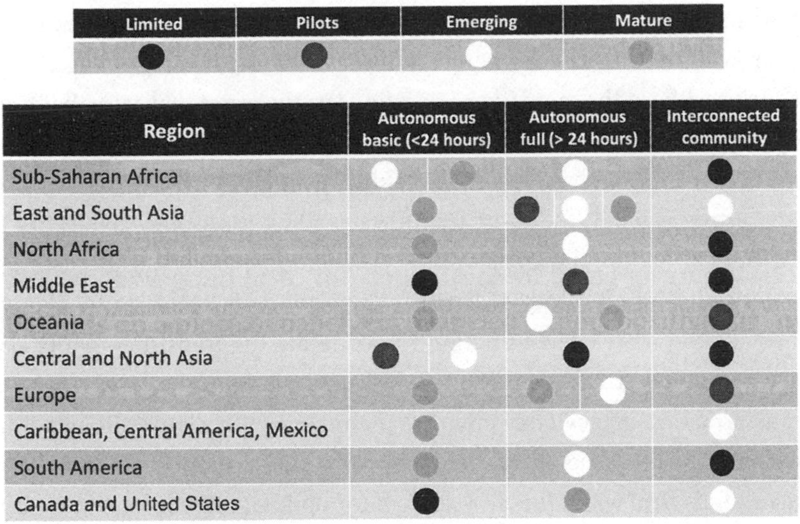

Limited	Pilots	Emerging	Mature
●	●	○	○

Region	Autonomous basic (<24 hours)	Autonomous full (> 24 hours)	Interconnected community
Sub-Saharan Africa	○ ●	○	●
East and South Asia	●	● ○	○
North Africa	●	○	●
Middle East	●	●	●
Oceania	●	○ ○	●
Central and North Asia	● ○	●	●
Europe	●	● ○	●
Caribbean, Central America, Mexico	●		○
South America	●	○	●
Canada and United States	●	●	○

FIGURE 4.3 Status of deployment for different types of mini-grids. *Limited* = very few markets; *Pilots* = isolated exploration; *Emerging* = developing markets; *Mature* = active deployment. *Autonomous basic* = systems for which power is supplied for less than 24 hours and may be turned off when there is insufficient renewable energy to meet load. *Autonomous full* = systems that can provide power on a 24-hour basis. *Interconnected community* = systems that may be used as a backup to the main grid, that are designed to sustain only the most critical loads, or that could be used to provide primary power, with the main grid as a backup. (International Renewable Energy Agency [IRENA], *Innovation Outlook: Renewable Mini-Grids* [Abu Dhabi, United Arab Emirates: IRENA, 2016], 18, figure DRE5, http://www.irena.org/DocumentDownloads/Publications/IRENA _Innovation_Outlook_Minigrids_2016.pdf.)

with both centralized grids and with other mini-grids.[47] Figure 4.3 assesses the evolution of the market for renewable energy mini-grids, ranging from limited and isolated exploration (pilots) to developing market (emerging) and active deployment (mature).

Still, established transmission and distribution systems remain the backbone of power grid infrastructure in most countries. Households, businesses, and other organizations rely on this essential public service to support their livelihoods. Safety, reliability, affordability, and environmental stewardship are still the core principles that guide decisions on electricity generation, transmission, and delivery. In this era of limited budgets, a transition needs to strike a balance between reinforcing these criteria underpinning energy security, while at the same time exploiting the synergies that arise thanks to digitalization and decentralization.

There is a risk with any groundbreaking technology that its introduction can destroy value in the short term by stranding assets and giving rise to transitional losses. The transition to DER could even result in customer defections from the grid. This risk compels energy planners to identify and take the most effective action to accelerate the transition, while also ensuring it is cost-effective. The World Economic Forum believes that this transformation is inevitable and that maintaining the status quo is not an option.[48] The key issues are therefore how the public and private sectors can successfully deal with DER and shape it for optimal impact.

The growth of DER presents opportunities to improve power systems; however, it also introduces key challenges. For example, ensuring the cybersecurity of the grid has become more complicated because of the development of the IoT, globalized supply chains, and an increase in small power generators.[49] DER also influences the conceptualization of service reliability. In addition to ensuring technological efficiency, grid managers must increasingly plan for contingencies such as generation loss because of extreme weather conditions. Sustained outages are becoming more common in this age of elevated storm activity. Affordability is a lingering concern, even in an era of declining or stable energy prices. For customers, the cost pressures of repairing and upgrading aging infrastructure are coupled with the cost pressures of incorporating "smart" technologies and DER initiatives into a preexisting, often technologically outmoded system while simultaneously ensuring that the system remains safe, cybersecure, and reliable.

In this vibrant era, countless new business ideas and technologies are competing for attention and investor or customer dollars. Not every new technology or potential resource can or should be adopted. Yet it is recognized that many DER initiatives can add real value to the operation of the electric grid, as well as benefiting customers and providers. Appropriate valuation of DER initiatives and smart grids are key challenges and essential building blocks for ensuring that they do in fact provide net value to the system.

A report by the World Economic Forum describes how countries could accelerate the deployment of new grid technologies and realize the economic and social benefits they deliver, including sustainability, security, reliability, and customer choice. The speed of adoption and the success in shaping the transformation depend on strategic choices regarding regulation, infrastructure, business models, and customer engagement. Those energy systems that fail to exploit synergies will fall short of capturing full value from DER. Traditional energy companies that fail to innovate risk the stranding of network assets and customer defections from the grid.[50]

Typically, technological adoption curves follow a logistic curve (s-curve) of adoption: adoption starts slowly, then accelerates, and then the rate of growth slows as saturation is approached (figure 4.4). In an increasingly interconnected world, these technology diffusion curves have become compressed; the speed at which nearly complete market penetration is achieved has shortened from 30 to 40 years for technologies such as the telephone and automobile to 15 to 20 years for technologies such as the VCR and the cellphone. There are indications that the benefits of clean energy technologies and smart grids will ensure that the current energy transition will be one of the fastest transitions humanity has seen. However, in the leadup, we need to figure out what to do with existing infrastructure, which might not be fully salvageable in a smart-grid world.

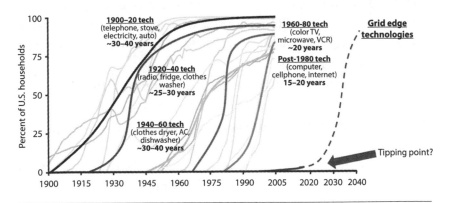

FIGURE 4.4 Time for technologies to reach 80 percent penetration (Alexandra May, *Power to the People: How the Sharing Economy Will Transform the Electricity Industry* [New York: World Economic Forum, 2017], https://www.weforum.org /press/2017/03/power-to-the-people-how-the-sharing-economy-will-transform -the-electricity-industry/.)

THE SMART DISTRIBUTED GRID

The electric grid in most industrialized countries is a large, burly thug of an artifact. It was designed to deliver electricity from large power plants via a high-voltage network to local electric distribution systems that then serve individual consumers. Like electrons on the grid, information was designed to flow predominantly in one direction, from generation and transmission to distribution systems through to consumers. A system of this type benefits from size. Therefore, most electricity grids around the world are run as monopolies. Historically, when drawing on conventional energy sources, a single firm has been able to produce all of the electricity needed in a city or even a nation at a lower cost than a collection of competing firms. Massive power plants with industrial boilers, turbines, generators, rail links to deliver coal, natural gas–pumping stations, cooling towers, and switchyards have been able to produce unbeatable scale economies.

But with the advancement of DER, the information and innovation promise of an increasingly digital society, growing threats to infrastructure security, and concern over global climate disruption, the centralized design with electrons and information flowing strictly from producers to consumers is no longer sufficient. By moving toward a smart distributed grid, numerous distributed assets can be integrated. Aggregators and intermediaries can begin to add supply-chain value, and consumers can better manage their electricity consumption, to the point of even helping to generate and balance grid loads.

Smart-grid architectures can integrate and optimize electricity coming from a diverse array of fuels and power generators, including centralized power plants as well as distributed renewable resources, energy storage, demand response systems, and electric vehicles.[51] This means that, if planned strategically, the smart grid can be a value-added enhancement rather than a costly replacement of existing infrastructure. We don't need to displace the burly thug; we merely need to refine it. Figure 4.5 portrays a complex smart-grid system with both central and regional controllers managing the two-way flow of electricity and information between utilities and consumers. The actual mix of controls and technologies will depend on a region's transmission and distribution system, its electricity governance

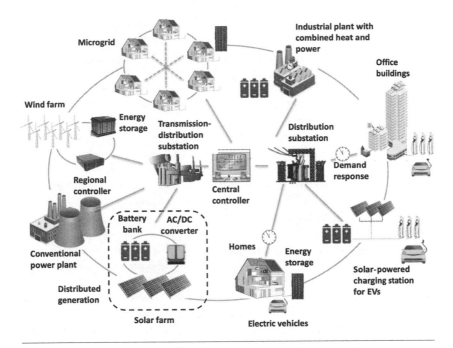

FIGURE 4.5 Smart grid, a vision for the future (Marilyn A. Brown, Shan Zhou, and Majid Ahmadi, "Smart-Grid Governance: An International Review of Evolving Policy Issues and Innovations," *Wiley Interdisciplinary Reviews [WIREs]: Energy and Environment* 7, no. 5 [2018]: e290.)

and business model, the nature of the customers being served, and other demand-side issues. In other words, this new technology is malleable.

By implementing a smart grid, electric systems can operate at higher levels of power quality and system security.[52] The efficiency of power delivery can be improved through dynamic pricing and flow control, which enable consumers to play an active role in managing their demand for electricity. Payment systems can be made more efficient with digital communications and can reduce nontechnical losses that undermine grid economics in many developing countries. Without the development of the smart grid, the full value of individual technologies such as distributed solar PVs, electric cars, demand-side management, and large central station renewables such as wind and solar farms cannot be fully realized.

Countries are at different stages of smart-grid deployment and face unique barriers and drivers for change (table 4.1). In the **United States**, smart-grid

TABLE 4.1 Energy policy drivers and policy emphases, by country

	Policy drivers	Policy emphases
United States	– Power system reliability – Clean electricity – Economic development and jobs	– Technical and operational standards – Smart meters – Dynamic pricing and demand response programs
Europe	– Renewable energy and energy efficiency – Improved productivity through innovation – Carbon emissions reduction	– Technical and operational standards – Competitive retail market – Smart meters – Transmission and distribution networks modernization
Japan	– Energy security – Carbon emissions reduction – Enhancing competitiveness of domestic industries	– Smart communities – Smart meters – Solar photovoltaic generation
South Korea	– Energy security – Enhancing competitiveness of domestic industries – Carbon emissions reduction	– Smart power grid – Smart consumers – Smart transportation – Smart renewables – Smart electricity services
China	– Reducing power generation disparities between regions – Reducing energy/carbon intensity – Strategic economic restructuring – Renewable energy	– Solar photovoltaic generation – Upgrading and modernizing urban and rural electric grid
India	– Economic growth and access to energy – Providing price signals to consumers – Renewable energy and electric vehicle integration – Reducing transmission and distribution losses	– Smart power grid – Electric vehicles – Quality power for all – Dynamic tariffs

Source: Marilyn A. Brown, Shan Zhou, and Majid Ahmadi, "Smart-Grid Governance: An International Review of Evolving Policy Issues and Innovations," *Wiley Interdisciplinary Reviews (WIREs): Energy and Environment* 7, no. 5 (2018): e290.

policies vary across utilities and states. Over the last decade, most states have implemented net metering policies and interconnection standards. The first net metering law was introduced in 1983 in Minnesota, and by November 2017, **forty** states had such policies.[53] Although net metering laws vary widely in content, most place limits on **distributed generation** (DG) solar. In about half the states there is a subscriber limit for DG solar, generally in the range of 1 **to** 5 percent of peak demand, and there is also a limit on the amount of power allowed in interconnection arrangements, generally below 1,000 kW.[54] Many U.S. utilities are installing smart meters using funds from the American Recovery and Reinvestment Act, and dynamic pricing programs are widely used in industrial and commercial sectors.

Smart-grid programs are critical components of the European Union (EU)'s low-carbon agenda. In the United Kingdom, British regulators have not only been very active in the rollout of smart meters and modernized distribution networks, but they have also supported innovation in low-carbon technologies. Japan and South Korea are both focusing on innovation and export of smart-grid technologies to strengthen the competitive advantages of their domestic industries. Power supply concerns stemming from the 2011 nuclear disaster in Japan also accelerated the country's investment in smart-grid infrastructure, with an aim to integrate variable energy sources. To illustrate the urgency behind this transition, in 2016, Bloomberg New Energy Finance predicted that Japan will install 6 to 10 million smart meters per year through 2022.[55] By mid-2017, Tokyo Electric Power Company (TEPCO) alone surpassed 10 million smart-meter installations.[56] China, the second-largest developing country in the world, sees smart-grid development as essential for renewable deployment and strategic energy industries.

Evidence from the past decade suggests that the rapid and widespread deployment of smart-grid technologies will not occur without supporting policies. A review of emerging smart-grid policies in the United States, EU, Japan, Korea, China, and India suggests that considerable progress has been made to develop effective policy frameworks. Nevertheless, further advances are needed to harmonize policies across nations, states, and localities and to learn from recent experiences with this new generation of electrical grid technologies.[57]

As interconnectivity is essential for large-scale, integrated deployment of smart grids, development of standards at the national and global level will be particularly important in the near future. Establishment of lead agencies to coordinate efforts at various levels of government would facilitate the standardization process and unify governance of cybersecurity challenges.

The electric power industry has been presented with a tremendous opportunity to create new revenue streams and to adopt a coordinating role in the emerging low-carbon economy. The costs required for the full deployment of smart grids are large, but so are the costs associated with upgrading dilapidated electricity grids. Currently, government is still the central driver in smart-grid investments. This fact suggests the need for a policy framework that attracts private capital investment, especially from renewable project developers and information and communication technology companies.

An electricity market that encourages competing business models could enhance the flexibility of electricity systems and support an increasing penetration of renewable generation technologies. Reforming the rate design mechanisms that are currently discouraging utilities' investments in advanced technologies and ensuring that costs and benefits are shared among all stakeholders are also important future challenges. Regulatory changes that remove barriers to a competitive energy market could also optimize overall operations and costs, hence increasing the net social benefits from smart grids.

As the deployment of smart grids progresses, systemwide optimization accompanied by progressive improvements in energy efficiency may significantly reduce peak demand and make some generation facilities redundant. In other words, if the electricity industry does not respond proactively to market dynamics, many firms might find themselves serving an evaporating customer base. A proactive response requires sophisticated resource planning and comprehensive cost-benefit analysis (CBA) at the early stages of smart-grid planning. Smart-grid customer policies, such as dynamic pricing and customer protection, require detailed understanding of customer behavior and adoption patterns. New policies need to be developed based on social science studies of consumer feedback and response to smart-grid technologies and regulations.

Collaborating on smart-grid standards and sharing experiences from demonstration projects can reduce repetition and overlap in smart-grid deployment efforts. Disseminating best practices can be particularly beneficial to those developing countries where electricity infrastructure is expanding rapidly and the scale of investment necessitates that strategies are implemented properly the first time.

Worldwide, there are several innovative smart-grid programs to learn from, including New York's Reforming the Energy Vision (REV) program, Italy's smart-meter program, and Korea's carbon-free Jeju Island initiative.

New York's REV Program

REV is an ambitious initiative of the governor and the Public Service Commission to fundamentally reorient the structure of the state's power sector from a centralized model to a highly decentralized one. The rationale for making this transformation is to "build a lasting market structure to support investment in, and the adoption of, clean energy at scale" by integrating DER into the planning and operation of electric distribution systems.[58] REV seeks to improve system reliability and resiliency, leverage systemwide efficiency, enhance customer knowledge, promote fuel and resource diversity, and reduce carbon emissions.

REV's success depends on the Distribution System Implementation Plan (DSIP), which in turn depends on competitive proposals for providing DER and utility distribution assets. The selection approach uses a benefit-cost analysis (BCA) that focuses on maximizing the benefits delivered by each component of the system. One lesson to learn from this approach is that fragmented BCAs of this type are not sufficient to achieve an optimal DSIP. Instead, an optimization approach that incorporates all systemic costs should be adopted to ensure the long-term success of this regulatory initiative.

The New York Public Service Commission elected to use a BCA to develop the DSIP, perhaps because of its apparent simplicity as well as regulatory familiarity and experience in the context of using this tool for energy efficiency program evaluation. The process aims to calculate a benefit-cost ratio for each proposal, to rank proposals from highest to lowest, and then to procure all proposals with a value above 1. A criticism of this process is that if the BCA includes nonmonetized benefits and costs, it is entirely possible that the cost to retail electricity consumers would be greater even when all of the accepted proposals have monetized benefits that exceed costs. To avoid excessive inflation of utility bills, the BCA ranking approach could be modified to adopt all measures for which the BCA exceeds 1 and the total cost is less than a prespecified amount (including requiring reductions) on utility bills. This approach would help ensure that utility costs are kept within an affordable range; however, selecting the highest benefit-cost ratio proposals that do not exceed the utility-bill-impact constraint does not ensure the most efficient outcome. The efficient use of society's resources (that is, maximizing social welfare) occurs by selecting the portfolio of proposals that maximize collective net present value. The ultimate lesson? Clearly, there are ideological choices that need to be made in evaluating the benefits of a smart-grid development plan, and transfer payments may be needed to avoid economically disadvantaging less affluent consumers.

Italy's Smart Grid

Italy emitted 467 million tonnes of CO_2 ($MtCO_2$) in 2006, but only 344 $MtCO_2$ in 2017. Its carbon intensity (at 5.3 metric tonnes of CO_2 per capita) is now lower than the EU average of 6.9, and much lower than the U.S. average of 15.8 metric tonnes of CO_2 per capita. Its diminishing carbon footprint is achieved in part by a significant investment in renewable power, which represents 43 percent of the nation's total electricity generation, eclipsing the EU average of 32 percent. Modernization and expansion of the electricity transmission and distribution networks have been a critical step in the successful integration of renewables in Italy's energy system.[59]

Efforts at various levels of government have been made to accelerate energy infrastructure optimization. In 2007, the European Commission approved the Operational Program—"Renewable Energy Sources and Energy Saving"—in southern Italian regions (Apulia, Campania, Calabria, and Sicily) with a budget of €1.6 billion (US$2.0 billion).[60] One priority of this program was to improve the infrastructure of transmission networks to accommodate renewable energy and small- or micro-scale cogeneration, which received €100 million (US$123 million) from European and Italian state funds.[61] Within this context, the Italian Ministry of Economic Development and Italy's largest power company—Enel Distribuzione (Enel)—together launched a €77 million (US$95 million) "Smart Medium Voltage Networks" project in southern Italy to make medium voltage distribution networks more amenable to PV systems with installed capacity between 100 kW and 1 MW.[62] In addition, the Italian Regulatory Authority for Electricity and Gas selected firms for eight tariff-based financial projects on active medium voltage networks, to demonstrate the at-scale advanced network management and automation solutions necessary to integrate DG systems.[63] Through these pilots, stakeholders learned how to optimize system integration, allowing other regions to benefit from learning by doing.

Italy has one of the largest and most extensive smart-meter programs in the world. In 2006, its legislators mandated the installation of smart meters in all low-voltage customers and specified minimum performance standards for the installed meters.[64] Italy's smart-meter deployment has an emphasis on distribution system integrity, designed to support the liberalization of the energy market and prevent electricity theft. Enel Distribuzione is the major player in Italy's smart-meter deployment program. Up to 2016, Enel had installed smart meters for 32 million customers within its Italian electrical distribution network and had provided advanced services enabled by smart meters, such as the introduction of an hourly based tariff system.[65] Enel has also extended its smart-meter system to its Endessa distribution grids in Spain, where 7 million smart meters were installed as of 2017. To encourage renewable DG development, the Italian government guarantees priority access to the grid for the electricity generated by renewables and provided feed-in tariffs to solar PVs and incentives for PV plants entering into operations between 2013 and 2016.[66] Incentives for new plants ended in 2017, as renewable technologies have effectively closed the cost gap on fossil fuel competitors.

Cybersecurity plays a critical role in Italy, the fourth largest market for information and communication technology (ICT) services in the EU. With operations in thirty countries, Enel has faced many cyberattacks.[67]

In response, Enel has implemented new technical standards: it participates in a joint initiative of four major European utilities to tackle cybersecurity, it has developed a "National Strategic Framework for Cyberspace Security" to protect the ICT infrastructure,[68] and the Italian privacy regulator Garante has issued six-step guidelines to protect data privacy.[69] For nations concerned about security issues, the Italian case yields valuable lessons.

South Korea's Carbon-Free Jeju Island Initiative

South Korea imports over 97 percent of the energy it consumes and is highly dependent on imported petroleum and liquefied natural gas. Its energy system emitted 688 $MtCO_2e$ (million tonnes of CO_2 equivalent emissions) in 2016, up from 579 $MtCO_2e$ in 2010 and 484 $MtCO_2e$ in 2006, mirroring its rapid economic growth, which is the fastest among Organization for Economic Cooperation and Development (OECD) countries.[70] Its per capita CO_2 emissions have also increased from 11.9 metric tons in 2010 to 13.8 metric tons in 2016, largely reflecting growing affluence and production growth in the steel sector. Renewable energy accounts for less than 2 percent of its electricity generation. If these trends continue, South Korea is poised to emit 728 to 744 $MtCO_2e$ in 2030 (147 to 153 percent above 1990 levels).[71]

As a non–Annex I party to the Kyoto Protocol, South Korea has not been obliged to reduce its carbon emissions. Nevertheless, the Korean government set a voluntary goal of reducing greenhouse gas emissions by 37 percent below its business-as-usual case by 2030 (equating to approximately 81 percent above 1990 levels, or approximately 536 $MtCO_2e$).[72] Reducing the nation's energy dependence and reducing its carbon intensity are two top priorities of Korean energy policy, in part because the government views energy efficiency improvements as a competitive enhancement. Accordingly, a mandatory cap-and-trade system has been operating since 2015.[73] But needless to say, the nation has its work cut out to meet its reduction targets.

The electric power system of South Korea is highly reliable and efficient.[74] Power generation, transmission, and distribution are managed by the Korea Electric Power Corporation (KEPCO), which has six power generation companies, four subsidiaries, and four affiliated companies.[75] KEPCO began the deployment of smart-grid technologies in 2005 when Korea launched a national program to develop digital, environmentally friendly, and intelligent electric power devices and systems and to advance Korean electric power and electrical industries.[76] In 2008, "Korea's National Strategy for Green Growth" was announced. It proposed an expenditure of 107 trillion won (US$101 billion) between 2009 and 2013, with investments focused on smart-grid technologies and smart cities.[77]

The associated energy blueprint—Smart Grid Road Map 2030—promotes investment in five sectors: smart power grid, smart consumption, smart transportation, smart renewables, and smart electricity services.[78] By 2030, a nationwide smart grid and 27,140 charging stations for electric vehicles are to be built, and smart meters and advanced metering infrastructure are to reach 100 percent penetration by 2020. In addition, South Korea plans to generate 11 percent of its energy from renewables and to achieve a maximum of 10 percent energy usage reduction by 2030. The plan aims to reduce the annual blackout time per household from 15 minutes in 2012 to 9 minutes in 2030, and it aims to decrease the power transmission and distribution loss rate from 3.9 percent in 2012 to 3.0 percent in 2030. A total of 27.5 trillion won (US$25.85 billion) is to be spent on technology development and infrastructure construction to implement the plan.

As a first step to implement the Road Map, the Korean government started a pilot program on Jeju Island in 2009, including a fully integrated smart-grid system for 6,000 households, wind farms for power, and four distribution lines.[79] A total of US$50 million in public funds and US$150 million in private funds was invested from 2009 to 2013. More than one hundred companies from automobile, renewable power, telecommunications, and home appliance sectors participated in the program. The next stage of the Road Map involved expansion into metropolitan areas. Between 2012 and 2016, policy goals and implementation strategies aimed to guide the construction of smart-grid pilot cities in seven districts.[80] The final stage will be development of a nationwide intelligent grid.

KEPCO is also developing export-ready nuclear power plants, electric vehicle charging infrastructure, integrated gasification combined cycle (IGCC) power plants, and carbon capture and storage (CCS) technologies.[81]

The anticipated benefits of implementing the Road Map include fifty thousand new jobs every year and 230 million tons of greenhouse gases avoided by 2030.[82] For nations looking to approach smart-grid development through staged rollouts, the Korean case provides insightful strategies.

COLLABORATIVE CONSUMPTION AND THE GREAT ENERGY TRANSITION

In many respects, smart grids can be vehicles of empowerment (pun intended). They empower firms to innovate and broaden product and service offerings. They also empower communities to take control of local energy futures through decentralized systems. But most significantly, they empower individuals—or at least, *can* empower individuals. To optimize the benefits from smart grids, the active participation of citizenry is needed. People need to be willing and prepared to make the investments and undertake the behavioral modifications necessary to tap the full potential of smart grids. A shift to a sharing economy should be considered an integral piece of the hardware for enabling smart grids that actually contribute to low-cost, low-carbon solutions.

Despite the fact that many from the Woodstock generation have traded in tie-dyed shirts and motorcycles for Boss suits and SUVs, the notion of a sharing economy continues to evolve. In 1978, sociologist Marcus Felson extolled the value of collaborative consumption as an alternative socio-behavioral paradigm.[83] Periodically since then, notions of sharing and peer-to-peer exchanges have quickly captured the imagination of analysts and practitioners.[84] The common denominator of these concepts is optimizing accessibility to services and utilizing products, rather than suffering the inefficiencies that ownership of assets implies.[85] The sharing economy essentially combines the essence of Woodstock and a craft fair at the highest levels of the commercial organization—the premise being that one can wear that tie-dyed shirt and still be able to justify driving a fuel-guzzling SUV.

Fundamentally, sharing is predicated on reciprocal benevolence. It is underpinned by the perceived willingness of many to share their property for a limited amount of time without any formal transfer of property rights.[86] Sharing in this light is seen as an alternative to private ownership, and it is facilitated by marketplace exchanges and through gift giving. "In sharing, two or more people may enjoy the benefits (or costs) that flow from possessing a thing."[87] Harnessing a spirit of benevolence and tapping the social benefits of sharing are seen as strong drivers of success and aspirational outcomes in the sharing economy.[88]

The rapid uptake of collaborative consumption arrangements suggests that the sharing economy invites interest and support across broad swaths of society.[89] Social media forums such as Twitter, Weibo, and Facebook amplify practices of sharing because so much more consumer information is being exchanged and connections are being rapidly made.[90] Increasingly, apps such as Airbnb and eBay serve as central markets where collaborative consumption reigns. In energy circles, the popularity of collaborative consumption parallels the shift toward a reconceptualization of energy to emphasize the services delivered by energy systems and to deemphasize the equipment through which these services are being provided.

Extremes of hyperconsumption and the environmental damage caused by exploitative practices have been credited as fueling interest in a sharing economy.[91] If assets are shared, the costs of superfluous consumption can be attenuated. Some have argued that sustainability can be better served by focusing on the satisfaction of using and experiencing goods and services rather than owning and having property rights to them all.

From an economic perspective, a core concept of the sharing economy movement is the ability to capture and redistribute the idle capacity of existing assets.[92] By increasing the usage rate of products and assets, the sharing economy can help to ensure that they are used to their full economic potential. In most advanced economies, owners drive their cars only a few hours each day, offices are often empty, most rooms in homes are unoccupied much of the time, stores have peak and off-peak shopping hours, and in conventional energy circles, "captive and cogeneration" power plants have substantial unutilized capacity. All the while, large numbers of consumers suffer from power shortages.[93] Collaborative consumption optimizes technical potential by putting this excess capacity to better use.[94]

Coincidentally, when existing capacity is better utilized, fewer raw materials and intermediate resources and products are needed to manufacture goods and services because increasing the usage of existing products reduces the need for more resources to make more products. Thus, while many view the sharing economy as important from an economic and business perspective, others laud the potential environmental benefits.[95]

The sharing economy also tends to be more localized, contributing to stronger communities. In our highly globalized economy, people depend on resources from outside their communities.[96] This dependence creates wealth for people outside their communities. By sharing assets and resources, people gain more localized benefits from consumption.[97]

The sharing economy can also take advantage of lower transactional costs compared to the overhead costs associated with buying.[98] It therefore offers a competitive advantage that can be exploited by startups or existing companies to turn a profit. Physical assets, infrastructure, skills, time, and land—all can be shared and monetized.[99] Thus, the sharing economy is a concept intertwined with economic value added.[100] All together, the sharing economy is a blueprint for a highly productive business model that shows how to synergize economic, environmental, and social issues.[101]

Rapidly falling costs of smart meters, connectable devices, and grid sensors will increase the efficiency of network management and, more importantly, allow customers to access real-time information about energy supply and demand across systems. This change will alter consumption norms significantly. Consider the following scenario. In a sharing economy, we might find a neighborhood where vehicles are all shared, thereby reducing the cost of vehicle access. Parking would be centralized, freeing up space that we used to call a home garage. The cars would charge during nonpeak hours by drawing from decentralized energy technologies erected in the neighborhood or from a smart grid, depending on which is cheapest at the time. At other times, batteries of idle cars would be accessed by households to top up energy supplies or store excess electricity. All of the benefits from owning these cars would be shared by the community.

In the same way that Uber or Airbnb have disrupted the transport and hospitality industries, respectively, smart distributed technologies could improve the utilization rate of costly electricity infrastructure. The electricity system was built to meet peak demand, meaning that, unlike President Trump's Twitter account, a significant portion of the infrastructure sits idle for most of the time. In the United States, the average utilization rate of the majority of generation infrastructure was below 55 percent in 2015. A decrease of 10 percent in peak demand could free up to US$80 billion of value by increasing the overall utilization rate of infrastructure.[102] This value could be harnessed by the rollout of smart distributed resources that are actively managed by customers. Under the right price signals and market design, customers would be able to produce their own electricity, store it and then consume it at a cheaper time, sell it back to the grid, or even share it with a neighbor.

These technologies and business models have reached their adoption tipping points, and both industry and regulators need to prepare for digitally connected, distributed resources. New York, Italy, and South Korea are already well down the path of experimentation. Globally, it is estimated that US$2.4 trillion of value will come from new job creation and the reduction of carbon emissions derived from increasing the efficiency of the overall system, optimizing capital allocation, and creating new services for customers.[103] If these benefits can be realized, the transition costs may be a drop in the ocean.

Yet, to be successful, numerous coping mechanisms are needed. First, the regulatory approach must be redesigned so that DER integration is facilitated and all benefits are monetized. Second, enabling infrastructures must be deployed to sire a dynamically evolving system that remains flexible, open, and interoperable, ensuring that customers and third parties can benefit from the data that they generate. Third, the customer experience must be redefined to afford more value-added benefits and to tailor customer experiences. Fourth and finally, new business models and financing schemes must be embraced to engender competition in the development of alternative and complementary goods and services equipped with new capabilities.

THE ROAD AHEAD

The scale and scope of the technical, socioeconomic, political, and business model uncertainties surrounding energy systems have never been higher. These have engendered fundamental questions about how societies wish to live, grow, connect, and sustain themselves.[104] Recently, the emergence of collaborative consumption has sired new players in many segments of the economy, motivated by the pursuit of better value generation across supply chains, greater environmental stewardship, technology advancement, and changing attitudes about product ownership and the need for social connection.[105] The energy sector is on the cusp of just such a transition. We need to ensure that the infrastructure is configured to foster this transition and that people are prepared to capitalize on the benefits to come.

Amid this backdrop emerges the question of how this expensive endeavor will be financed in a world where politicians seem intent on amassing levels of financial debt that paralyze investment in major works projects. The next chapter takes up this challenge.

5

Fostering and Financing the Energy Infrastructure Transition

I n 2008, the global economy suffered the worst financial crisis since the Great Depression of the 1930s. It began in 2007 with the crash of the U.S. subprime home mortgage market, and it spiraled into a full-blown international banking crisis by 2008. The GDP of most North American, European, and Asian countries dropped by more than 10 percent, with energy consumption following suit. What happened next was unprecedented. By 2017, the global economy had rebounded to its 2006 level, but energy consumption and greenhouse gas (GHG) emission levels barely budged in many countries around the world. There was a global decoupling of energy consumption from GDP.

This decoupling was enabled by a significant shift in how society consumes energy. Over the last decade, energy-efficient technologies and smart energy practices have better penetrated the marketplace. At the same time, many advanced economies have continued to shift to high-tech service economies, and cleaner fuels gained market share.[1] This move to lower-carbon fuels has amplified the CO_2 reductions derived from energy efficiency. In sum, a stealthy transition to energy efficiency is under way.

One consequence of decoupling energy usage from economic growth is that the energy to power homes, businesses, mobility, and manufacturing is now more affordable. Yet there are undoubtedly more cost-effective energy efficiency and decarbonization initiatives waiting for the right champions.[2] Advocates have long considered that energy efficiency is the least-cost energy resource; some even argue that it could enable a carbon mitigation pathway with "negative costs."[3] But this notion of reducing GHG emissions while saving consumers money violates many preconceptions within the school of neoclassical economics. Although markets for energy efficiency technologies are expanding and evidence is increasing that energy savings on the customer side of the meter are large and, in many cases, profitable, there are those who continue to deny that there is an "energy efficiency gap." These Milton Friedman groupies argue that all truly cost-effective efficiency investments have already been made and that further energy efficiency would come at a net cost to society.[4] Writing in 2008, as the economic downturn was unfolding and an era of energy efficiency was about to begin, one skeptic claimed that energy efficiency improvements are marginal because consumers would simply rather spend their money elsewhere than invest in an energy-efficient initiative.[5] As we look ahead at options for fostering and financing the energy infrastructure transition, we will be reiterating this staggering level of market myopia that is alive and kicking in many markets.

While chapter 6 will more comprehensively describe the array of policy instruments that can affect change and enhance innovation, this chapter focuses primarily on ways to foster and finance an extreme energy infrastructure makeover. We begin by exploring the current state of energy research and development (R&D) with special emphasis on a promising global endeavor known as Mission Innovation. These R&D efforts will then be placed in proper context when we examine the expected investment requirements for fostering a full transition and consider strategies that can be enacted to enlist the financial participation of a broader swath of stakeholders. One thing is clear from our analysis: all hands will be needed on deck to spur the pace of transition. The global total annual investment in the energy sector is estimated to be over US$1.2 trillion.[6] But clearly more will be needed if this energy *Titanic* is to change its course. We elaborate on strategies to enhance the impact of investments and the role that carbon finance can play in mitigating risk. The chapter closes with reflections on alternative financial models that are emerging in both developed and developing nations and that hold promise of further effectiveness.

ENERGY INNOVATION AND THE GLOBALIZATION OF CLEAN ENERGY R&D

R&D is a powerful engine of change, and the payoff from R&D that is focused on advancing energy technology is well documented. But there is underinvestment in research on advanced energy systems, as well as on climate adaptation and resilience. Why is this? Research points to many common actions that provide collective benefits that are not fully rewarded in the marketplace; as a result, they are routinely undersupported.

In 2010, the National Academy of Sciences concluded that the availability of advanced energy technologies could decrease the cost of CO_2 emission reductions from US$80 per tonne of CO_2 (tCO_2) to perhaps $25/$tCO_2$ in the United States by 2020 and could deliver further cost reductions on the road to meeting mid-century GHG emission reduction goals.[7] Given this possibility, a key challenge is how to craft policy to overcome stakeholder reluctance to invest. One way to encourage and direct private-sector R&D investments is for governments to initiate public-private R&D partnerships with technology road-mapping, collaborative development, and cost sharing.[8] These are features common to two notable international R&D initiatives: Mission Innovation and Breakthrough Energy Ventures.

Mission Innovation was launched in 2015 as countries around the world came together to discuss the Paris Climate Accord and accelerate the development of tomorrow's clean energy technologies. Mission Innovation is a global initiative of twenty-two nations and the European Union, representing about 60 percent of the world's population, 70 percent of global GDP, and more than 80 percent of government investment in clean energy research.[9] The goal is to significantly accelerate the pace of innovation, reduce costs, and make clean energy technologies affordable worldwide. Member nations have committed to double their public investments in clean energy R&D

over five years while encouraging collaboration among partner nations and between businesses and investors. The program has established itself as a global rallying point for advancing clean energy research and innovation.[10]

In June 2016, all Mission Innovation member nations adopted a collective baseline of contributing US$15 billion per year in clean energy R&D investment. The members have further pledged to work toward a doubling in publicly directed clean energy R&D investment over a five-year period, to reach a combined US$30 billion per year by 2020–2021.[11] Between 2016 and 2021, it is estimated that this investment would inject an additional US$25 to US$35 billion in clean energy R&D, on top of the business-as-usual investment projections of about US$75 billion.

Figure 5.1 shows the accumulation of baseline commitments and projected energy R&D expenditures of Mission Innovation member countries. While the United States is forecast to spend US$6.4 billion in both baseline and forecast years, China's energy R&D spending is expected to more than double, from US$3.8 to US$8.9 billion, by 2021. These Mission

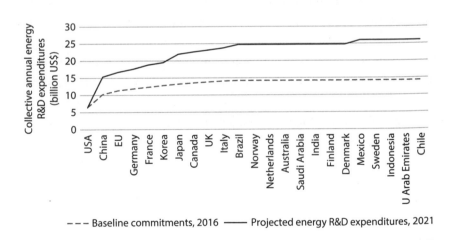

FIGURE 5.1 Annual and projected R&D expenditures of Mission Innovation members (Redrawn from data published in Z. Myslikova, K. S. Gallagher, and F. Zhang, *Mission Innovation 2.0: Recommendations for the Second Mission Innovation Ministerial in Beijing, China*, CIERP Climate Policy Lab Discussion Paper, Tufts University, 2017, https://sites.tufts.edu/cierp/files/2017/09/CPL_MissionInnovation014_052317v2low.pdf.)

Innovation efforts represent an unprecedented acceleration of R&D efforts for advancing clean energy technologies.[12] Member countries have identified seven innovation challenges to guide R&D investments: smart grids, off-grid access to electricity, carbon capture, sustainable biofuels, advanced solar conversion, clean energy materials, and affordable heating and cooling of buildings.

While each Mission Innovation member organizes its own clean energy R&D portfolio to serve national priorities, members may also pursue collaborations in areas of shared interest to leverage complementary assets and further accelerate progress. Collaborative R&D workshops are held to bring together experts from different scientific disciplines so that they can identify and prioritize clean energy areas ripe for further investigation and explore collaborative endeavors that can benefit from broader-scale scientific research.[13] The Mission Innovation workshop handbook recommends using bibliometrics (the statistical analysis of bibliographies) to help with workshop planning, to identify patterns and trends in the field of research, and to assess the potential network strengths of individual researchers, institutes, and countries. For example, figure 5.2 portrays the networks connecting the top thirty research universities and institutes, based on

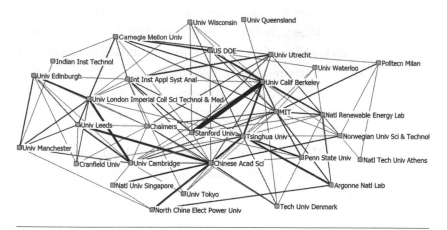

FIGURE 5.2 Relationships of coauthored publications on low-carbon electricity, 1990–2016 (Lu Wang, Yi-Ming Wei, and M. A. Brown, "Global Transition to Low-Carbon Electricity: A Bibliometric Analysis," *Applied Energy* 205 [2017]: 57–68, http://dx.doi.org/10.1016/j.apenergy.2017.07.107.)

contributions to publications on low-carbon electricity.[14] It shows that scientific research on low-carbon electricity is a highly collaborative endeavor. In some cases, the collaborations involve the close coupling of neighboring domestic institutions (e.g., University of California-Berkeley and Stanford in the United States and Imperial College and Cambridge University in the United Kingdom). In China, where knowledge transfer is vital to progress, the Chinese Academy of Science and Tsinghua University form hubs for engendering international collaboration.

Mission Innovation recognizes the importance of science and early-stage R&D as the drivers of truly transformative breakthroughs in energy innovation. Use-inspired energy research can nurture novel approaches to project development and feed the innovation pipeline to attract more applied public and private R&D investment. This is the kind of investment that can lead to breakthroughs in scientific understanding and problem solving. Early-stage research activities are characteristically pre-commercial, so they may provide greater opportunities for intergovernmental cooperation and represent a natural fit for scientific exchanges and information sharing across international boundaries. Given the sensitivities about international competitiveness, publicly supported activities in late-stage R&D on near-commercial technologies often entail involvement by the private market sector, which frowns on open sharing of insights. International activities at the near-commercialization stage are more sensitive to competitive concerns and tend to be carefully governed to protect intellectual property.

Created at about the same time, Breakthrough Energy Ventures (led by Bill Gates, Mark Zuckerberg, and twenty-seven other billionaire entrepreneurs) complements Mission Innovation by encouraging private investment in low-carbon energy innovation. By the end of 2017, Breakthrough Energy Ventures had raised US$1.3 billion and was seeking to fund game-changing startups that can deliver reliable, affordable, low-carbon products to the world.[15] This approach builds on evidence suggesting that redirecting small fractions of revenues to R&D for a sustained period of time can result in breakthroughs that dwarf initial research outlays.[16]

Where is clean energy R&D and innovation occurring? Figure 5.3 illustrates the spatial and temporal distribution of publications centering on clean energy technologies. With an average annual growth of 50 percent

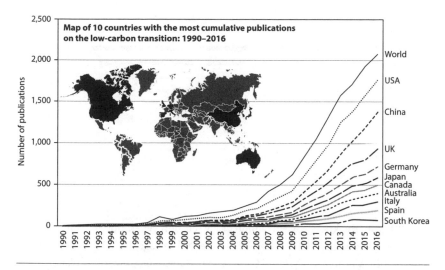

FIGURE 5.3 Exponential growth of publications on electricity decarbonization, 1990–2016 (Redrawn using data from Lu Wang, Yi-Ming Wei, and M. A. Brown, "Global Transition to Low-Carbon Electricity: A Bibliometric Analysis," *Applied Energy* 205 [2017]: 57–68, http://dx.doi.org/10.1016/j.apenergy.2017.07.107.)

over the past two decades, academic efforts in clean energy R&D are clearly expanding. The signing of the 1997 Kyoto Protocol was precursor to a growth spurt in 1998, and the 2015 Paris Accord may have amplified the growth of Chinese publications in recent years. Overall, published research results tend to be dominated by researchers in developed countries, but China stands out as the emerging star of the developing world. In 2016, China overtook the United States to become the most productive country in terms of publications on clean energy. China's ascendance is bolstered by national electricity sector reforms and the launch of a national carbon market on December 19, 2017. Together researchers from China (21 percent) and the United States (19 percent) account for approximately 40 percent of the world's published R&D articles. An additional 40 percent of published R&D articles come from the United Kingdom (10 percent), Germany (6 percent), Japan (5 percent), and Canada (5 percent), as well as from Australia, Italy, Spain, and South Korea (all with 4 percent).[17] Increasingly, developing countries are playing an important role in clean-tech publishing, challenging the long-held domination of developed countries.

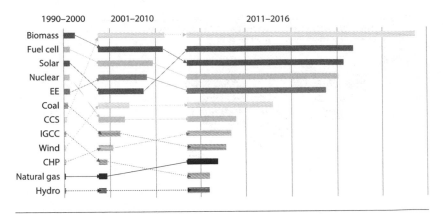

FIGURE 5.4 The knowledge evolution of low-carbon generation technologies; EE = energy efficiency, CCS = carbon capture and storage, IGCC = Integrated Gasification Combined Cycle, CHP = combined heat and power (Lu Wang, Yi-Ming Wei, and M. A. Brown, "Global Transition to Low-Carbon Electricity: A Bibliometric Analysis," *Applied Energy* 205 [2017]: 57–68, http://dx.doi.org/10.1016/j.apenergy.2017.07.107.)

As developing countries begin to address climate change adaptation, resiliency, and environmental protection, more opportunities to exploit distinct expertise, knowledge, and other resources are emerging between developed and developing countries.[18]

Not only is the geography of clean energy R&D evolving, but so is its focus. The technological direction of the global transition to low-carbon electricity is evident through bibliometric analysis and visualization. Figure 5.4 illustrates the evolution of R&D focus by ranking the comparative amount of research on twelve energy technologies over time. This analysis uses publications related to a given technology as a proxy for highlighting where efforts are being directed.

Since the 1990s, energy research has substantially increased, mirroring the enhanced global awareness of climate change and environmental protection. Carbon capture and sequestration (CCS) has experienced the most significant ascent, capturing first place as the most researched technology. Solar, fuel cells, and energy efficiency research round out the top five.

Within these technological categories, there are a number of substreams. For example, biomass includes forest biomass; woody, cellulosic, and lignocellulosic biomass; microbial and algal biomass; and palm oil and biodiesel. Solar research includes solar photovoltaic (PV) panels, solar thermal assemblies, solar power towers, concentrating solar collectors, utility-scale solar farms, and distributed rooftop systems (with or without battery storage). CCS research centers on the pre- and post-combustion capture of carbon and exhibits a strong focus on cost-benefit analysis. Wind technologies continue to attract researcher attention, advancing from ninth to sixth place over the past two decades and dominated by research into offshore wind. While conventional coal and natural gas still serve as key generation resources, there has been a relative decline in publications associated with associated breakthroughs and new technologies—the exception being CCS research.

The evolution of technological research tends to mirror market trends, and themes tend to reflect engineering challenges during a given period. The reliance on baseload coal and nuclear fuels in the 1990s incentivized the search for cleaner alternatives, resulting in the rise of CCS and wind in the first decade of the twenty-first century and the growth of solar and natural gas (including combined heat and power [CHP]) in the second decade. Recently, there has been an increase in articles related to green electricity policy analysis and life cycle assessment, reflecting the emergent challenge of expediting the energy transition amid large-scale opposition. There has also been an evolution toward more nuanced forms of economic and policy analysis.[19] Despite our hopes that this book will become the go-to book for energy transitions, this avenue of research will likely evolve further as competing perspectives congeal.[20] Regardless of which supply-side technologies emerge as dominant, a steady stream of research into smart technologies and energy efficiency initiatives is likely to continue to grow, reflecting the favorable political winds that support research in these areas.[21]

THE INVESTMENT COST OF THE GREAT ENERGY TRANSITION

Estimating the direction, size, and timing of future investment flows is a precarious undertaking, and few have engaged in it using rigorous, well-resourced analysis. We'll begin this section by examining the various scenarios for a clean energy future that academics and think tanks have produced, as one way to frame the investment costs of the great energy transition. We then turn to the prognostications of energy consultants who provide industry with well-researched and experience-based overviews of current energy investment trends. These two sources serve to book-end the cost spectrum, ranging from low to medium transition costs.

It should be intuitively obvious that cost drives strategy when it comes to energy transitions. Explicating the financial requirements of energy transition is important for policymakers and the business community, and it tends to underpin negotiation positions when it comes to designing future climate change mitigation strategies.[22] Understanding the cost of alternative energy futures is also important to NGOs, energy consumers, and other stakeholders whose views and decisions will influence infrastructure investments—where to live and work, what education and career paths to follow, what capital investments to make, whom to lobby, whom to vote for, and how much insurance to buy. Fortunately, the research community is beginning to pay more attention to these issues.

For the first time, the Fifth Assessment Report (AR5) of the IPCC Working Group III dedicated a whole chapter to examining the crosscutting investment and financial implications of climate change mitigation policies.[23] "Total climate finance" is examined, including "all financial flows whose expected effect is to reduce net GHG emissions or to enhance resilience to the impacts of climate variability and the projected climate change." This covers private and public investments and considers the full value of the financial flows, not just the share associated with the climate change benefit (e.g., the entire investment in a wind turbine).[24]

Using this framework, total investment flows were estimated to range from US$343 to US$385 billion per year between 2010 and 2012, with the vast majority of investment (about 95 percent) directed at emission reduction efforts. Between US$45 and US$120 billion was concentrated in developing countries (including public climate finance of US$35 and US$49 billion), and international private climate finance flowing to developing countries ranged from US$10 to US$72 billion per year (including foreign direct investment as equity and loans). Problematically, during the same

TABLE 5.1 Global annual investment shifts from 2010 to 2029 to limit warming to 2°C

	Reduction in global investment	Increase in global investment
Investment in fossil fuel–fired power plants without CCS	– US$30 billion	
Investment in low-emissions generation technologies (renewable, nuclear, and electricity generation with CCS*)		US$147 billion
Investment in energy efficiency in the building, transport, and industry sectors, involving modernization of existing equipment		US$336 billion
Investment in reduction of fossil fuel extraction, transformation, and transportation with higher energy efficiency and a shift to low-emission energy sources	– US$116 billion	
Net financial change	– US$146 billion	US$483 billion

Source: Derived from data published in S. Gupta et al., "Cross-cutting Investment and Finance Issues," in *Climate Change 2014: Mitigation of Climate Change. Contribution of Working Group III to the Fifth Assessment Report of the Intergovernmental Panel on Climate Change* (Cambridge: Cambridge University Press, 2014), 1211.
*CCS = carbon capture and sequestration.

period (2010–2012), global CO_2 emissions from fossil fuel energy increased by almost 5 percent.[25] Clearly, then, much more is required on the financial front to alter course.

To limit warming to 2°C, investment must be significantly expanded and redirected. The IPCC AR5 describes a scenario, shown in table 5.1, that outlines the financial costs of requisite annual investment reallocations during the period 2010–2029. Even with investment in fossil fuel activities rechanneled to clean energy, this analysis suggests that investors will need to muster US$337 billion per year to foster the needed infrastructure and energy efficiency initiatives.

The private sector is already playing an important role in financing mitigation efforts. Its contribution was estimated at US$224 to US$267 billion per year in 2010–2012, accounting for between 62 and 74 percent of overall climate change mitigation finance.[26] Since energy costs are such large expenses in most countries, this is not surprising. Nevertheless, private capital investment still has room to grow under stronger enabling environments—more supportive financial institutions, competition-leveling regulations, and tighter guidelines to ensure the security of property rights. In many leading countries, a large share of private-sector investment in climate change mitigation also relies on low-interest loans and risk guarantees from governmental agencies. This fact again underscores the role of the public sector as a facilitator of private investments. Given that leaders in most nations are hard-pressed to redirect fiscal funds at the scale required to sustain this transition, a strong argument can be made that private financing will likely continue to be central to spurring on the transition.

The IPCC projection for financing investment in support of climate change mitigation (US$337 billion) can be compared to current trends in the industry to see if there is alignment. According to a REN21 report, since 2010, global new investment in renewable energy and new fuel has averaged US$273 billion, with US$280 billion flowing into the sector in 2017.[27] So there is work to be done. Energy industry experts tend to have a good grasp of current trends and access to financing. According to Tim Rockell, director of Global Energy Institute at KPMG, an energy grid capacity buildout across all fuels is currently under way that if unaltered will see capacity expand over sixfold by 2050.[28] As a result of this rapid electrification trend, for the first time, total investment going to the electric sector (traditional power and renewables) is surpassing the investment in oil and gas.[29]

Current trends suggest that coal and gas will continue to play an important role going forward, but their commercial fates appear to be moving in different directions. Increasingly, it is difficult to find financing for coal projects because banks are shying away from such investments. On the other hand, although natural gas projects also face political scrutiny, the role that natural gas can play as a peak-load technology in support of renewable energy development insulates natural gas from the same fate as coal.

Natural gas is projected to be the largest source going forward, but renewables are gaining momentum.[30] Over the next few decades, experts forecast that renewables will account for 20 to 25 percent of electricity generation, with solar, wind, geothermal, and mini-hydro (run-of-river) dominating. This rapid growth of renewables is partly a function of their economically palatable size, as 30 to 100 MW projects are much easier to finance than mega projects are. Barring new developments in storage technology, natural gas will have a role to play in helping utilities accommodate increasingly high amounts of stochastic energy flows from renewable technologies. In other words, renewable technologies have found a "frenemy"—friend and enemy—in natural gas and vice versa.

In addition to challenges associated with accommodating amplified loads from renewable energy sources, land acquisition is proving to be a constraint on growth for large-scale solar and wind farms in some regions. In response, small-scale microgrids are emerging as attractive alternatives, and this development is changing the financing requirements. Through community financing mechanisms and microfinance strategies, remote areas of the world are gaining access to electricity in a way that keeps investment dollars in a given community. As a result, energy access is expanding worldwide. Nevertheless, the price tag is not inconsequential: bringing modern cooking and lighting to people now relying on rudimentary sources could yield substantial improvements to human welfare but would require US$72 to US$95 billion per year until the job is done in 2030.[31] This is a lot of microfinance to coordinate.

The assessment of investment flows in the power sector under different climate mitigation scenarios is in its infancy. Few studies have used integrated assessment models on either a global or a regional scale.[32] The shift toward a smart grid with distributed resources and consumers acting as prosumers is much more complicated to model that the traditional centralized system of large-scale power plants, which deliver unidirectional flows of electricity (and information). The scope of energy investment is becoming much more diverse. While there will be room and, in many nations, preferences for traditional large-scale projects driven by big energy, there is also a burgeoning role for cooperative investment and small-scale private investment such as what we have seen with wind power development in Denmark or community solar projects in India. Given such complexity, how do we estimate the costs of all this?

First, let's consider what we know about financing from a U.S. context, and then we will turn to investments in China. The point being, if both China and the United States are investing heavily in renewable energy technologies and other climate mitigation technologies, similar investments will almost be irresistible for other nations that will not want to lose ground to these two economic behemoths.

In the United States, during the Obama administration, the EPA and others estimated that implementation of a low-carbon pathway, particularly the Clean Power Plan (CPP), would cost several billion dollars more in the year 2030 than the cost of "business-as-usual." However, when energy efficiency initiatives are added to the policy package, the mitigation cost estimates in 2030 amount to several billion dollars less than "business-as-usual."[33] Specifically, the United States could save money and comply with the CPP if strong energy efficiency policies are adopted to promote investments in better-quality construction, industrial upgrades, window retrofits, enhanced insulation, wider adoption of smart thermostats, LED lighting, and high-efficiency air conditioning and boilers, to name but a few examples. Extending this analysis to include deeper CO_2 cuts by 2040 (a 56.6 percent reduction in 2040 relative to the reference case) amplifies potential savings. Quantitatively speaking, without energy efficiency initiatives, meeting cumulative CO_2 reduction targets from 2016 to 2040 would require a cumulative incremental investment of US$529 billion (in 2011US$), but when energy efficiency efforts intensify, the cumulative incremental investment is reduced to $159 billion.[34] In short, comprehensive systemic thinking is required to turn costs into benefits.[35]

China has become the leading investor in renewables. In 2016, it invested more than any other country in solar PV, concentrated solar power (CSP), and wind and solar water heating; and according to a REN21 report, the total renewable capacity in the country at the end of 2017 stood at 334 GW, a 29 percent increase from 258 GW in 2016.[36] More impressively, by 2020, China aims to install a further 340 GW of hydropower capacity, 110 GW of solar power capacity, and 210 GW of wind power capacity. Between 2015 and 2017, China invested US$344 billion in new renewable power and fuels, and in 2017, Chinese renewable power investment amounted to 45 percent of the global total.[37] Although this scale of buildout is costing China a fortune, energy system expansion is essential to its continued prosperity and to a reduction of the fossil fuel pollution that plagues its major cities. The good news is that as the nation's economy grows, so too does the capacity of the nation to finance further investment.[38]

Large-scale human needs often must be met through networks composed of large-scale infrastructure systems, which in turn sculpt the geography of a community, region, or nation. Consider the impact of the electric power grid; the transportation network of rails, roads, and highways; the systems for water collection, distribution, and treatment; and the great assortment of buildings that together house people as they undertake economic activities in cities. These infrastructure systems are ubiquitous and critical to modern societies; they embody and depend upon knowledge, strategies, leadership, norms, and a specialized workforce. To excel, infrastructure networks of this type are designed, built, and maintained within silos that promote efficiency. Yet there is evidence that optimization of resources requires society to optimize system-wide integration of its infrastructure investments.[39]

A good example of a forward-thinking approach to planning infrastructure is the Tennessee Valley Authority's (TVA) US$300 million strategic initiative, which will expand its fiber capacity to improve the reliability and resilience of the transmission system while adding further enhancements to support local economic development. Piggybacking on transmission tower rights-of-way, the network expansion will help meet the power system's growing need for bandwidth as well as accommodate the integration of new, distributed energy resources. With a modernized fiber backbone of this type, TVA can free up spare fiber capacity to help local communities and rural areas develop high-tech economies and attract and retain jobs, because broadband connectivity is vital for business and industry. The fiber initiative will take perhaps five years to complete and will require 3,500 miles of fiber cabling.[40]

Society's failure to exploit the interconnections among infrastructure systems has resulted in lost opportunities to exploit cross-sectoral savings and benefits. In energy, this suboptimal buildout has been tolerable only because of the availability of cheap fossil fuels and because the externalization of costs, risks, and harms is passed on to the general public through taxation. The lack of infrastructure resiliency has been exposed in the United States by natural disasters (e.g., the Superstorm Sandy crisis), economic corruption (e.g., the subprime mortgage crisis), and social forces (e.g., the Flint, Michigan, water crisis)[41] that are likely to become more frequent and severe in an era of changing climate, resource scarcity, and institutional transparency, connectivity, and accountability.

CARBON PRICING AND MARKET-BASED FINANCE

If this were a world of competitive climate markets and strong ecosystems of suppliers, it's possible to imagine that large-scale mobilization of financing might meet mitigation and adaptation needs.[42] Carbon taxes and the auctioning of carbon allowances generate public revenues that can be used to foster investments in clean energy technologies. There is already evidence of this strategy. In 2010–2011, for example, carbon taxes generated about US$7 billion in revenue annually, mainly in countries participating in the European Union's emissions trading system (EU-ETS). According to a World Bank estimate, these revenues are set to grow as more aggressive policies come online.[43] Currently, regional, national, and subnational carbon pricing schemes cover 22 percent of global CO_2 emissions (figure 5.5).[44]

The problem is that collecting taxes is just the first stage of an optimized solution for fostering a clean energy transition. Ideally, these tax revenues should then be used to facilitate capacity expansion and enhanced R&D into clean energy. Unfortunately, revenues from carbon pricing schemes can be spent in many different ways. For example, none of the EU-ETS revenues have been earmarked for international climate finance initiatives.[45] The Carbon Dividend Plan in the United States recommends returning revenues from carbon taxes to households on a per capita basis to facilitate revenue neutrality rather than growing the coffers of centralized governments.[46] However, such a plan does little to direct investments into transitional energy initiatives. Often the use of tax revenues becomes a political issue, and so a compromise solution might be the only viable alternative.

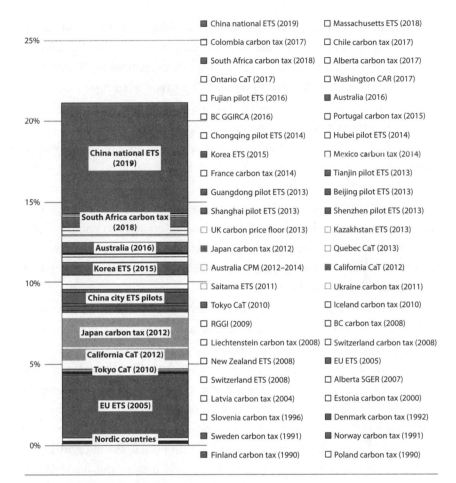

FIGURE 5.5 Regional, national, and subnational carbon pricing initiatives: share of global emissions covered; ETS = emissions trading systems, CaT = carbon tax, BC GGIRCA = British Columbia Greenhouse Gas Industrial Reporting and Control Act, CPM = carbon pricing mechanism, RGGI = Regional Greenhouse Gas Inititiative (United States), CAR = Clean Air Rule, SGER = Specified Gas Emitters Regulation (Drawn by authors from data published in World Bank and Ecofys, *Carbon Pricing Watch 2017* [Washington, DC: World Bank, 2017], https://openknowledge.worldbank.org/handle/10986/26565.)

For example, for the Regional Greenhouse Gas Initiative in the northeastern United States, some tax revenues from the program can be used to develop and deploy clean energy technologies while others are used to eliminate the distorting effects of other fiscal policies.[47] This type of compromise is one way to enhance an energy transition through the development of coalitions that also wish to tap into these tax revenues.

Carbon pricing is widely seen as the least-cost economy-wide policy to reduce CO_2 emissions because it standardizes the marginal cost of abatement across diverse sources, technologies, and consumers.[48] Based on Pigouvian taxation principles, once the carbon tax is set (e.g., per ton of CO_2 emissions), it offers flexibility by allowing emitters to choose abatement strategies.[49] Carbon pricing also makes it possible for governments to adjust tax levels over time as marginal social damages from climate change become stronger or weaker.

In contrast, placing regulations on polluters or on specific abatement technologies can distort markets in unintended ways. Regulators, policymakers, and the public that elects them inevitably have imperfect information about how to optimize regulations.[50] In particular, regulators might not know what technologies to support. For example, if a given nation wishes to support expansion of wind power, its policymakers also need to ensure that the pace and scale of expansion can be supported by grid peak-load capacity.[51] If stochastic energy flows are not adequately supported, the integrity of the system begins to break down, giving rise to added costs and the threat of system failure. As a result, direct regulation might not yield effective and sustainable long-term solutions.[52]

Perhaps the most significant weakness of regulations is an inability to catalyze transformative change. Regulations represent the cudgels of the policy tool kit; one either meets them or not. Under a regulatory policy, a power plant that fails to meet a regulation runs the risk of being shut down. Conversely, if the power plant is being taxed on emissions, the cost of failing to reduce emissions simply makes the plant less profitable.

Similarly, regulations do not effectively incentivize demand-side change. Initiatives such as promoting end-use energy efficiency, population control, and the minimization of food waste are all worthy initiatives when it comes to abating CO_2 emissions.[53] It would be a bold government that decides to regulate behavior in these areas. In short, conceptually, taxes incentivize change, whereas regulations compel change.

With that said, carbon taxes can be imperfect, too. They create winners and losers and can give rise to unintended consequences. For example, carbon taxes might encourage municipalities to phase out carbon-intensive power plants, but these taxes do not dictate specific market direction. A community that shutters a coal-fired power plant because of high taxes might wind up—as many communities in China have discovered—with nuclear power plants in their backyards. Carbon taxes also require a great deal of information to ensure that tax levels are set optimally: too high and social welfare can suffer; too low and targets of the tax are not motivated to act.[54] Because lower-income households use a larger share of their earnings to buy electricity and gasoline, the taxation process can be regressive and create inequitable outcomes by penalizing poorer consumers.[55] Carbon taxes can also be ineffective if adjoining jurisdictions fail to implement or enforce them, resulting in leakage and spillovers, as occurred in the United States with the Regional Greenhouse Gas Initiative.[56] Also, while carbon taxes provide certainty about the cost of mitigation, they provide limited assurance about the resulting level of emission reductions.

Carbon caps—the process of limiting CO_2 emissions at the source and then charging the emitter for any amount of emissions that exceed the cap—provide more certainty than carbon taxes alone, because caps when aggregated establish specific limits that must be achieved. However, this mechanism can amplify costs because caps need to be allocated to specific emitters and the process needs tighter enforcement. Leniency can be built into the process so that a cap resembles a tax if a mechanism is introduced to allow low emitters to sell their savings to excessive emitters through the auspices of a carbon market. This is the essence of a cap-and-trade system.

Another twist on the carbon cap strategy is to auction off allowances rather than simply assign allowances. As with carbon taxes, the revenue from auctioned allowances in cap-and-trade programs can help decrease the regressive financial burden of emissions reduction efforts.[57] Moreover, the tax revenues stay in the system, and those who need to purchase additional allowances can do so from those who managed to keep their emissions under their allocated allowances.

There is increasing awareness that implementation of these new carbon policy instruments—carbon taxes, caps, cap-and-trade, and tradable allowances—can help to address the policy uncertainty associated with energy investments, which is an important barrier to the domestic deployment of low-carbon technologies. In principle, both a carbon tax and a carbon cap mechanism will provide policy stability.[58]

For these reasons, carbon market strategies are springing up like politicians at a church picnic. Carbon taxes have been used in five northern European countries since the early 1990s (Denmark, Finland, the Netherlands, Norway, and Sweden). In 2001, the United Kingdom followed suit by implementing a climate change levy (CCL), which was applied to the industrial, commercial, agricultural, public, and service sectors. The revenues raised by this tax were used to offset other taxes, making this policy revenue-neutral. Carbon taxes have been established in several U.S. jurisdictions, including the San Francisco Bay area and Boulder, Colorado.[59] They are also used in other regions of the world, including British Columbia, Canada, and Australia. The British Columbia tax started at $10/tCO$_2$ in 2008 and increases by $5/tCO$_2$ each year until eventually hitting $50/tCO$_2$ in 2021. All of the tax revenues are redistributed through corporate tax cuts, personal income tax cuts, and low-income tax credits. The resulting revenue neutrality potentially creates a strong double dividend if the funds are re-tasked for clean energy investments.[60]

On November 4, 2016, the Paris Agreement entered into force, marking a new era of emergent growth for carbon pricing instruments. As of July 13, 2018, 194 parties had signed the agreement and 179 parties—representing 88.75 percent of global GHG emissions—had deposited their instruments of ratification.[61] As part of the agreement, all ratifying nations have to develop and publish GHG emission reduction strategies. Unsurprisingly, given the benefits of the carbon policy instruments just outlined, two-thirds of the nations state that they are considering using carbon-pricing schemes directed at reducing 58 percent of global GHG emissions.[62] As of 2017, over 40 national and 25 subnational jurisdictions have put a price on carbon in some form or another.[63] According to the World Bank,[64] carbon pricing mechanisms around the world helped governments raise US$52 billion in 2017.[65] As is the case with rain in Seattle, more is expected.

As a sign of things to come, China and regional clusters of states (such as Canada, Mexico, etc.) are gearing up emission trading systems in the coming year. China's system will be the largest carbon pricing system in the world and is of particular importance because of the message that it sends to other nations, particularly the United States. Given that these two nations together account for over 40 percent of global emissions, unless both are firmly on board, any global transition away from carbon-based energy forms will stall.

Momentum is also building in support of carbon pricing in the private sector, where an increasing number of companies are actively attempting to manage climate-related risks. According to the environmental disclosure firm CDP, the number of companies that reported use of an internal price on carbon in 2016 more than tripled compared to 2014. Over 1,200 companies—including more than 100 Fortune Global 500 companies with a total annual revenue of about US$7 trillion—disclosed to CDP in 2016 that they are currently using an internal price on carbon or plan to do so within the next two years.[66] Given that over 7,000 companies with over 50 percent of global market capitalization disclosed environmental data through CDP in 2018, the 2016 CDP carbon price adoption rate of 17 percent suggests that there is more to come. This is particularly salient following the 2017 recommendations of the Financial Stability Board (FSB) Task Force on Climate-Related Financial Disclosures, which advised companies to disclose climate-related financial risks and opportunities and to disclose the internal carbon prices used.[67]

COMPLEMENTARY FINANCING OPTIONS TO FOSTER THE TRANSITION IN DEVELOPED COUNTRIES

The World Bank may be correct, and international climate markets may gradually take the place of financial assistance programs.[68] But in the interim, assistance programs, subsidies, and other policies continue to enable specified types of low-carbon technology investments and influence the evolution of energy-efficient products and practices.[69]

In addition to having to compete on an uneven playing field, clean-technology firms often are hindered by many other market hurdles, including principal-agent problems, imperfect and asymmetric information, and regulations that reward consumption over conservation. These hurdles are not addressed through carbon taxes. Consequently, carbon pricing is often discussed in tandem with policies that attempt to attenuate these barriers. Addressing many of the problems that hinder investments in energy efficiency has been found to be particularly cost-effective.[70]

Although government and market support for improving energy efficiency and greater renewable energy capacity is on the rise, a dearth of available and affordable financing remains one of the most significant barriers to expediting a clean energy transition. Investment capital became a scarce resource following the global financial crisis in 2008; while it is only now rebounding, it is still not sufficient to meet global requirements.

Fiscal policies and financing programs can help overcome these liquidity constraints, and the Paris Accord has spawned fresh government commitments, coupled with a virtual cornucopia of financial support schemes. For example, new programs in the United States allow firms and consumers to tie repayment of investments in energy efficiency and renewable energy to taxes or utility bills, allowing investors to offset costs through operational savings rather than as capital expenditures. Every state in the United States offers financial incentives to overcome the high-upfront cost of individual energy-efficient appliances and equipment, for both businesses and homeowners.[71]

Tax rebates and deductions are also common, as well as scalable to match program objectives. Always quick to innovate, California now scales some of its rebates, offering one rebate for measures that save at least 15 percent of a home's energy use, and a larger rebate for larger savings.[72] Italy offers one of the world's highest tax deductions for building retrofits. After four years of operation, the program has attracted €8 billion of investment by taxpayers and has resulted in annual energy savings of 4,400 GWh.[73] German homeowners can borrow up to €75,000 to invest in a more efficient home that may also include low-carbon power generators such as fuel cells or solar cells.[74] In the United States, the Northwest Ductless Heat Pump Project illustrates the effective bundling of rebates and tax credits with information programs. Over one hundred public and investor-owned utilities funded this project, which is designed to replace inefficient electric resistance heaters in existing homes and to overcome customer apathy, lack of training and education, and high upfront costs.[75]

In addition to subsidies, governments and private entities have experimented with a range of financing programs that pair limited public investment with high levels of private leveraging. Each of these approaches has unique advantages and disadvantages, and they all result in expanded access to financing through leveraging.

Clean-tech advocates view on-bill financing as one of the most effective financial programs to support renewable and energy efficiency investments. On-bill financing typically involves an investment made by a utility company for solar installations or energy efficiency improvements in a building, which is then repaid by the owners through their monthly utility bills.[76] While these programs require the participation of a utility, they also often benefit from the support of local governments in terms of legal authority and financing. If planned well, such programs give rise to added revenue streams by way of financial service provision and, as part of the utility bill, pose little to no hardship for the customer because energy bill savings typically outweigh upfront costs. In the United States, on-bill financing programs are expanding. Thirteen states have on-bill financing programs, and another five states have pilot programs or are in the process of creating programs.[77] In addition, the U.S. Department of Agriculture is now making on-bill financing available to small rural cooperatives. However, it does require a supportive utility to manage the program or engage a third-party contractor.

Another unique approach to funding energy efficiency projects is the use of private contractors to identify savings. Under an energy savings performance contracting (ESPC) agreement, a building owner contracts with an energy service company (ESCO) that finances and is responsible for retrofitting a building or facility. In some cases, this contract includes a guarantee of savings to the property owner, thereby eliminating the owner's risks. In virtually all cases, the project risks are borne by the ESCO, which is repaid through a share of the measured and verified energy savings (the building owner also typically accrues some of the savings over the life of the project).[78] In Europe, ESCOs work at a large scale thanks to government interest and support. There are many examples of ESPCs underpinning large projects involving municipalities, universities, schools, and hospitals.[79] As many European cases attest, governments that support such programs through regulatory, technical, and financial assistance mitigate risks and amplify buy-in. Government oversight can also prevent ESCOs from focusing only on the lowest hanging fruit and missing the full suite of economically attractive efficiency investments, which can become lost opportunities.[80]

Other government programs are more conventional in nature, offering tax incentives for investments. Consider what has happened with financing of distributed solar projects in the United States. Rather than enabling distributed solar systems via on-bill financing, some utilities responded to the declining cost of PV solar by investing in utility-scale solar farms. However, ratepayers were anxious to own their own solar panels and began to pressure state-level politicians for support. Many states responded with tax-based financial incentives, epitomized by the property assessed clean energy (PACE) financing scheme.[81]

Under PACE, state governments have authorized financing programs to facilitate distributed generation solar and energy efficiency projects by allowing building owners to pay for the costs of installation through an annual property tax payment. PACE financing allows property owners to finance rooftop solar panels and energy efficiency upgrades through assessments on their real estate tax bill, benefiting now and paying later. Special assessments are used to secure municipal bonds that raise the capital needed to support the PACE projects. Property owners' debt is repaid through the property taxes collected by municipal governments, typically over an extended period of fifteen to twenty years.[82] Since 2008, twenty-eight U.S. states and the District of Columbia have passed legislation enabling the creation of PACE districts.[83]

Of all the policies employed by governments around the world to promote renewable energy, feed-in tariffs (FITs) remain the most common, and the IPCC considers them to be the most successful.[84] FITs are programs that provide guaranteed grid access over an extended period (typically fifteen to twenty years), with prices based on the cost of generation plus a reasonable rate of return. The FIT payment is usually administered by the utility, which in turn is compensated either through government transfers or through permission to levy an additional charge for electricity. By the end of 2017, ninety-eight countries, states, and provinces had FITs, up from thirty-four in 2004.[85]

Other financing policies for renewable energy include production tax credits, investment tax credits, loan guarantees, direct public investment in infrastructure, low-interest loans, project grants, and renewable energy credits associated with renewable portfolio standards. In all cases, the overall premise is the same: subtly reduce the cost of a renewable energy project to encourage private investors to make the investment. In many cases, the level of support needed is not a lot. For instance, the U.S. production tax credit (PTC) in support of renewable energy development is a corporate tax credit that subsidizes selected renewable electricity generation. A PTC is provided to firms that generate renewable electricity (and nuclear and most recently CHP). In return, the tax burden for the firm is reduced (by 1.1¢ to 2.3¢ per kWh generated), generally for ten years. In the United States, renewable energy fortunes tend to mirror the trajectory of the PTC. In some years the PTC has been in danger of not being renewed, and this has stopped renewable energy diffusion in its tracks.

OVERCOMING FINANCIAL INVESTMENT BARRIERS IN DEVELOPING COUNTRIES

In the absence of strong enabling environments—robust financial institutions, regulations and guidelines to ensure the security of property rights—NGOs have traditionally taken up some of the slack when it comes to supporting clean energy markets in developing countries. They have contributed in multiple ways and have historically been key activists in shaping policy and impacting public opinion. They have also invented novel strategies to promote mass deployment of green technologies to resource-constrained local communities. For example, in Kenya and Uganda, NGOs have been influential in providing renewable energy infrastructure[86] and upgrading cookstoves.[87] In highly practical ways, NGOs are providing access to clean technologies in countries and rural areas where public entities and market will are weak or absent.[88]

From a financial perspective, NGOs have also become influential financiers in developing countries by serving as catalysts for microfinancing, one of the preferred modes of financial assistance to local communities in the developing world. Microfinancing entails the creation of groups within communities that can assist poverty-stricken villages by lending people (most often women) money without the requirement of credit or collateral for seeding small businesses. Grameen Shakti provides an example of this approach in Bangladesh, promoting solar panels, biogas plants, and improved cookstoves. It engages communities in projects that are delivered at the household and village levels and also engages regional and national policymakers as well as international donors. Using this approach, Muhammad Yunus and Grameen Bank have shown that even the poorest of the poor can work to bring about their own development.[89]

In the developing world, home cooking and heating are often done by burning wood, charcoal, and dung. These fuel sources generate substantial GHG emissions and produce environmental pollution that causes serious health hazards. This problem has engendered efforts to bring cleaner, more efficient stoves to impoverished villages, particularly in Africa, where poverty is greatest. Through microfinancing and international aid, new cooking and heating technologies are being introduced to those in need. Research on cookstove projects in Kenya suggests that carbon finance can also help build a vibrant market by attracting international actors and technologies, helping establish standards for monitoring installations, and facilitating better follow-up and after-sales support.[90] NGOs such as the Global Alliance for Clean Cookstoves, Practical Action, and Energy 4 Impact are leading the way in terms of promoting off-grid access to cleaner sources of household energy.

Another developmental need in emerging economies is in the area of electrification. About 1.3 billion people worldwide do not have home access to electricity, and the vast majority of these people live in rural areas. To deliver much-needed energy to these regions, the Total Access to Energy program was launched in 2010 by, ironically, a consortium of oil and gas companies to test and develop new business models. Within this program, the Awango project focused on the sale of solar-powered lanterns that can also charge cellphones. The project involved tailored financial support ranging from microloans to partnerships with NGOs. It also engaged in tangible, collaborative interactions with local communities, aiming to develop a sales strategy with realistic profitability targets that enable the project to grow and address its social mission of reinvesting profits in the project. By May 2015, the project was deployed in over thirty countries, and 1 million lamps were already sold, providing benefits to 5 million people.[91]

The Energy and Resources Institute—a leading think tank dedicated to conducting research for sustainable development in India and the Global South—also launched an ambitious program to promote solar lighting, called "Lighting a Billion Lives."[92] Typically, a solar charging station is set up in a village and is then used to charge the solar lanterns, which households can then rent for specific time periods at pre-set rates. Over time, the program has expanded to include other technologies, such as solar panels, mobile chargers, cookstoves, and LED light bulbs. As of 2018, the program had expanded to thirteen countries and more than 1.13 million households had been illuminated.[93] The project brings together key stakeholders—the technology provider, financial institutions, NGOs, energy enterprises, and the end users (the households).[94] Different forms of financial support have been used to facilitate the program, including funding from corporate and government bodies and support from international aid agencies for rural banking infrastructure. Studying after nightfall and not having to travel long distances to charge one's phone are two of the many life-changing benefits that can result from these types of programs. They also stimulate new economic activity and grow businesses.

In recognition of the financial limitations of many NGOs, Super ESCOs are now playing expanding roles in developing countries. These are energy service companies established by government agencies with the goal of implementing projects in public facilities and facilitating the development of private ESCOs. For example, the Indian Super ESCO, which is a quasi-governmental entity, uses multiple service providers to enhance its reach. ESCOs are also active in Croatia, the Philippines, and South Korea.[95] While they are still at a nascent stage in China, the Chinese government has been especially supportive of this financing approach as a means of scaling up energy efficiency improvements.[96] The province of Hebei, for example, aims to establish a 600 MW energy efficiency power plant, implemented by the Fakai Scientific Electricity Services Limited Corporation and using the Super ESCO model. The company is both developing and implementing projects using the ESPC approach, as well as assisting other ESCO operations in Hebei to grow their businesses and undertake more ESPC projects.[97] After years of development, Dubai now has an established, comprehensive regulatory system for supporting ESCOs, and several companies have gained accreditation.[98]

CONCLUDING THOUGHTS

Many novel business models and financing platforms, national programs, and various policies are emerging to help citizens in developing and underdeveloped nations to improve their lives through better access to clean energy. However, the process, as it stands, is agonizingly slow. Moreover, as lamentable as it is to mention, improving the lot of impoverished individuals does not help to attenuate GHG emissions unless they sidestep the fossil fuel pathway and go directly to renewables. As a case in point, millions in China have enjoyed the benefits of modern cookstoves, but they have also enjoyed elevated levels of affluence that have enabled them to purchase cars and other energy-consuming devices. Consequently, oil and coal consumption in the nation continues to rise.

Clean-tech financing approaches are clearly enablers of a low-carbon future, but innovation and stakeholder coordination are needed to ensure that a transition occurs at a fast enough pace to help impoverished nations leap-frog into a cleaner economy while also encouraging people in developed nations to eliminate waste and improve how resources are used. The next chapter takes up this challenge: How do we foster the innovation and coordinated response necessary to avoid the worst of what climate change has to offer? As will become apparent, our kindergarten teachers were indeed correct when they told us that collaboration is a good thing. For the challenge at hand, the collaborative task is to better understand what needs to be changed and design a suite of integrated policies that will promote desired change. Accordingly, the next chapter will continue the discussion on applied policy by examining the types of policy instruments that are being used to move markets and alter behavior.

6

Policies for Driving Innovation and Expediting the Transition

I n 2007, journalist Tom Friedman chastised China for its apathetic approach to renewable energy, noting that if China doesn't want to move to renewable energy—fine. From an American economic perspective, he argued, that will give the United States a chance to establish market leadership and enhance its competitive advantage.[1] Meanwhile, across the Pacific, in the lead-up to the Intergovernmental Panel on Climate Change (IPCC) Copenhagen Summit in 2009, China's then premier Wen Jaiobao explained China's perspective on an energy transition: "Action on climate change must be taken within the framework of sustainable development and should in no way compromise the efforts of developing countries to get rid of poverty."[2]

The premise underpinning Friedman's perspective was that strong domestic policies seed strong firms. Wind power examples include Vestas in Denmark, Siemens in Germany, and Gamesa in Spain. In 2009, the United States was poised to take on a leadership role in the area of clean energy while China was still adding 1.2 GW of coal-fired power each week.[3]

How quickly the world can turn. Today, China looks to capitalize on the clean energy economy as the U.S. government retreats from it. China now leads the world in installed solar power and hydropower capacity; and, after surpassing the United States in installed wind power capacity in 2010, it not only now boasts over twice the installed wind power capacity as the United States, but it currently hosts over a third of all installed wind power capacity.[4] It is clearly stepping forward to fill the leadership vacuum left by President Trump's dismissal of climate change. Rather than trying to apply new technology to reopen its abandoned coal mines, China is deploying its depleted open-pit coal mines as reservoirs that host large floating solar projects. The provincial government in Liulong, China, where coal has been king, hopes to collectively produce as much power from floating solar panels at retired coal mines as from a full-size commercial nuclear reactor.[5] In contrast, the U.S. Department of Energy (DOE) issued a Notice of Proposed Rulemaking in 2017 that would extend the lifetime operation of coal-fired power plants in the United States.[6]

Similar flip-flops in "fuels du jour"[7] have reversed the fate of countries that were once clean-tech leaders. For example, Japan was a world leader in solar panel production, largely on the back of its household solar thermal promotion programs. However, when the policies expired and the government committed to an escalation of nuclear power development (prior to Fukushima) as the central cog in its greenhouse gas (GHG) emission reduction efforts, the solar panel industry was usurped by firms in Taiwan and South Korea, and later on in China. This all tells us that the fortunes of energy technologies and the aspirations of energy policymakers are subject to changes in political will that are as fickle as an overfed cat. Yet for all the reasons explicated in earlier chapters, there is ample reason for a nation to get behind the coming energy transition. It will give rise to new industries, new specializations, more employment, and offshoot innovations. So how can such changes be catalyzed?

This chapter focuses on policy instruments as strategic mechanisms for expediting change. After examining policy rationales, theory, and logic, we turn to the choice of policy instrument. We first examine the rationale for governments to intervene before we employ a well-used policy tool framework to describe the types of policies that have been utilized to influence behavior in the energy sector. We conclude by highlighting the merits of adopting a portfolio approach to policy planning.

WHY GOVERNMENTS SHOULD INTERVENE

Most neoclassical economists subscribe to the market failure theory of public policy, which justifies policy interventions only when markets create socially suboptimal outcomes. In well-functioning markets, the demand and supply of a commodity such as energy reach an equilibrium that is determined by the price of a good, which in turn, in a competitive market, reflects the marginal costs of each technological cluster. Under this mindset, since energy is a relatively homogeneous product, the technology with the lowest cost becomes the market leader and the leader wannabe firms must endeavor to improve their products to compete.

Implicit in the validation of this worldview is the acceptance of rational actor theory: consumers have access to complete information and they base decisions on optimizing personal utility. With utility-maximizing consumers and perfectly operating markets, there are no problems that would justify government action.[8] In many capital cities, neoclassical economists strut around like pigeons in a park of popcorn vendors, influencing policymakers into believing that markets will logically work things out on their own.

But the world is not always fair and just. Most energy markets are replete with imperfections. Seven of the more prominent imperfections are illustrated in table 6.1.[9] These forces inhibit investment in clean energy technologies and even stop cheaper technologies and money-saving ideas from gaining market share.[10]

Consequently, for very valid reasons, governments need to get involved in altering market and consumer behavior. As we saw in earlier chapters, there is evidence that fossil fuel interests intentionally create misinformation campaigns to preserve market shares, so information is not so perfect.[11] Clean energy firms face huge investment barriers, and consumers must incur "switching costs" for moving from old to new technology, so transaction costs in energy can be high. We live in a busy time, and people are bombarded by information, so economists have invented a new term to describe consumers apathy—*rational inattention*. Moreover, even if most people had access to complete information for making decisions, they might not have the capacity to do so because there are so many moving parts involved in the complex, adaptive system that is the energy world.

TABLE 6.1 Types of imperfections in clean energy markets

Type	Illustrations in Clean Energy Markets
Imperfect and asymmetric information	Principal-agent relationships give rise to asymmetric information: landlords do not understand the needs of leasers, builders may undervalue life cycle costs, and energy consumers might not even know that they are wasting money.
High transaction costs	Markets are not frictionless. It takes financial and human resources to participate in markets. Advanced energy equipment may not be in stock, and trained personnel are often unavailable. Siting of wind farms can take months, if not years, of community stakeholder engagement.
Limited cognitive abilities	Individuals and firms are limited and biased in their evaluation of energy alternatives. They prefer avoiding losses to acquiring gains and are influenced by habits, which advantage incumbents and slow the pace of change.
Imperfect competition	Regulations in many nations give private electric utilities monopoly power. The OPEC oil cartel manipulates oil prices, making alternative liquid fuels less attractive. Patents may be blocked or not upheld. For years, fossil fuel technologies have received the lion's share of government subsidies.
External costs and benefits	Numerous unaccounted-for environmental and climate externalities occur in the mining, processing, and utilization of fossil fuels. The energy security benefits of distributed energy resources go largely unpriced.
Excludability	Principal-agent problems abound, in which users are afforded little say in the energy technologies that they have access to. Examples include renters who must use equipment that landlords provide. Investment and switching costs for new energy technologies can also exclude lower-socioeconomic-stratum members. For example, carbon taxes can hurt individuals who cannot afford to replace their ineffienct technologies, and gasoline taxes punish those who must, for financial reasons, commute long distances to work.
Limits to monetization	Some damages are caused by economic activities that cannot be adequately accounted for under a purely economic paradigm. For example, what price would you place on losing a loved one from a nuclear plant meltdown or on the extinction of species as the result of climate change?

Sources: B. K. Sovacool, M. A. Brown, and S. Valentine, *Fact and Fiction in Global Energy Policy* (Baltimore, MD: Johns Hopkins University Press, 2016), table 3.1; Marilyn A. Brown and Yu Wang, "Energy-Efficiency Skeptics and Advocates: The Debate Heats Up as the Stakes Rise," *Energy Efficiency* 10, no. 5 (2017): 1155–1173, table 1, http://link.springer.com/article/10.1007/s12053-017-9511-x.

As detailed in chapter 2, fossil fuel interests have received billions in government support over the years, and they now enjoy the spoils of their subsidized investments, so the competitive playing field is in dire need of leveling. Meanwhile, globally, the extraction of energy resources such as coal and uranium produces pollution that is rarely (bordering on never) incorporated into the cost of doing business.[12] In policy language, external costs are not adequately internalized into the price of energy. With over 7 billion consumers on the planet, we are now seeing the emergence of global environmental problems that are proving to be highly intractable.

Governments respond to these challenges by choosing from an array of behavior-altering tools to change market dynamics. In many ways, this is precisely what public policy is all about—altering behavior to address emergent needs. These behavior-altering tools are referred to as *policy instruments*. Because there are so many ways to influence behavior, academics categorize policy instruments and discuss their use in different contexts. One of the better-known frameworks is Christopher Hood's NATO framework, which is an acronym for nodality (information and messaging), authority (regulations), treasure (financial incentives and market-based approaches) and organization (government programs or agencies).[13] Figure 6.1 provides a few prominent examples of energy policy initiatives couched within the NATO framework. We will update and apply this framework to describe the various policy instruments that are in service around the world to catalyze a transition to a low-energy footprint.

One common denominator guides the selection of policy instruments. The behavior that a policymaker is trying to modify takes place in a contextually infused environment. Therefore, for policies to be successful, they have to be designed in line with the complexities of the contextual environment. This fact suggests that one must know the players and be clear on final goals before designing any policy intervention. In the immortal words of Yogi Bera, "If you don't know where you're going, you'll end up somewhere else."[14]

Common Nodality Instruments

Nodality instruments are interventions that attempt to alter behavior by informing people or coercing change.[15] This approach can entail providing

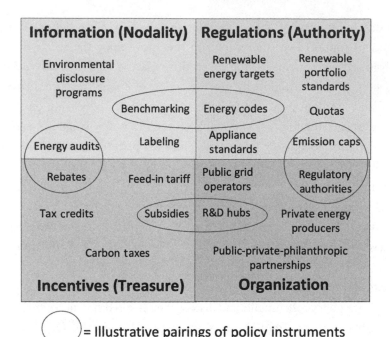

Information (Nodality)		Regulations (Authority)	
Environmental disclosure programs		Renewable energy targets	Renewable portfolio standards
	Benchmarking	Energy codes	Quotas
Energy audits	Labeling	Appliance standards	Emission caps
Rebates	Feed-in tariff	Public grid operators	Regulatory authorities
Tax credits	Subsidies	R&D hubs	Private energy producers
Carbon taxes		Public-private-philanthropic partnerships	
Incentives (Treasure)		**Organization**	

◯ = Illustrative pairings of policy instruments

FIGURE 6.1 Examples of different types of energy policy instruments (Authors' diagram.)

simple information (e.g., FAQs on climate change), reminders (e.g., switch off the lights when leaving the room), instructional guides (e.g., how to lower your energy costs), or messages intended to alert people to the existence of other government policies (e.g., feed-in tariff FAQs) or government programs (e.g., free energy audits).

The allure of nodality instruments is that these interventions are inexpensive and, if prepared in a targeted fashion, can be influential at nudging change along. The weakness of most nodality instruments is that people are free to ignore these messages. Since there is no driver to mandate change (penalties, financial incentives, or regulations), nodality instruments lack punch and are rarely used in isolation. Most often nodality instruments form part of a portfolio of policy instruments.[16] This is especially true in today's technological climate, where social media and the Internet can be strategically harnessed to get messages to the general population in a low-cost manner and can also be used to target specific segments of the population. In the following paragraphs, we introduce some of the more prevalent nodality tools that are used for facilitating energy efficiency and a clean energy transition.

Energy labeling of products encourages consumers and addresses knowledge gaps, providing much-needed information to help consumers understand the true costs associated with their technological choices. These labels are commonly used to "nudge" industry forward, unlike the more "muscular" treasure-based approaches taken to encourage energy efficiency and renewable energy in homes and businesses.[17]

Although a bit outdated, a report commissioned by the Australian Department of Industry indicated that 81 nations possessed product energy labeling programs in 2013, up from 50 nations in 2004.[18] One of the earliest transnational programs appeared in 1992, when the European Commission initiated a framework for developing product-specific energy labeling in Europe.[19] Today, most white goods, light bulbs, and cars must have an EU Energy Label displayed when being sold or rented. Energy efficiency is rated from A (the most efficient) to G (the least efficient). Products in Europe often also sport eco-labels, used as part of a voluntary scheme that promote products with high levels of environmental performance.[20]

There are energy labeling programs in all leading energy-consuming nations, including Canada, European Union (EU) nations, Australia, New Zealand, Japan, Singapore, and Taiwan.[21] In the United States, the Energy Star label is known by a majority of U.S. consumers, particularly for white goods. Surveys of those who recently bought cooling and heating appliances and products indicate that around 65 percent of the purchased products qualify as Energy Star approved, and about 85 percent said that the Energy Star rating played a significant role in the decision of which product to purchase.[22] The China Energy Label (CEL) Program, which was introduced in 2005, now mandates labeling in over twenty-five product groups, including motors, air conditioners, refrigerators, washing machines, gas kettles, water kettles, photocopiers, air compressors, flat-screen televisions, and fluorescent tubes.[23] The Chinese labeling program illustrates the transnational benefits of such programs. For many Chinese manufacturing firms, the cost of meeting the highest domestic energy efficiency standards is compensated by the fact that many of these firms are already producing energy-efficient products to meet overseas energy labeling requirements.

The success of these programs depends on the host country's ability to enforce compliance. In Australia, for example, one-third of the models claiming Energy Star compliance failed in verification tests.[24] With the growth in voluntary certifications around the globe, such shortfalls in compliance are increasingly an issue. In response, the European Ecodesign Directive has tightened requirements for verification.[25]

Another nodality instrument is *energy benchmarking* of energy use in homes, businesses, and industry. Such programs have the potential to transform markets by reducing information asymmetries in the marketplace and to enhance longer-term investment by consumers because they have greater knowledge of the savings to be had.[26] For example, in Singapore, residential energy consumers receive monthly utility bills that compare the consumer's energy consumption with that of other residents in their neighborhoods. In industries, mandated disclosure and benchmarking programs have been used for many years in Europe, and the United States is now beginning to implement them. Many U.S. cities now require the disclosure of energy and water consumption, as in the New York City Benchmarking Law, which requires large buildings (over 50,000 square feet) to report consumption data in a standard format using EPA's Portfolio Manager.[27] By understanding the scale of energy and water expenditures within a given building, tenants can make more informed choices when comparing rental costs.

In New York, the plaNYC plan released in 2007 to decrease the city's carbon footprint included energy performance reporting. By 2010 and 2011, data had been reported for over 2,000 buildings. The 2010 report suggested that if the bottom performers of large commercial buildings could be brought to just the median level of energy performance, energy consumption from this building class would fall by 18 percent.[28] By 2012, over 15,313 properties were subject to the underlying benchmarking ordinance, with 84 percent compliance.[29] This example highlights just how quickly these types of programs can produce concrete results and just how important it is for nodality tools to be combined with other types of instruments to enhance their impact.

Numerous studies have shown higher occupancy rates, higher rents, and higher property values for high-efficiency buildings.[30] Therefore, benchmarking helps enhance market value for energy-efficient buildings. Research further suggests that offering financial assistance in tandem with benchmarking would also help improve effectiveness.[31] In 2017, Atlanta's energy benchmarking program was buttressed by the addition of property assessed clean energy (PACE) financing that is repayable over the long term via property taxes.[32]

Industrial-level benchmarking is an area with substantial potential. Because of the uniqueness of industrial processes and intellectual property (IP) concerns, information on how energy is used by individual manufacturing enterprises is scarce.[33] At the top of the list of International Energy Agency (IEA)–recommended energy efficiency policies for industry is an appeal for access to high-quality energy consumption data.[34] Such data can be used to identify and motivate "best in class" performance in energy efficiency.

There has been some progress in this regard. Benchmarking is used in several countries to compare energy use among different industrial facilities within similar sectors.[35] In the Netherlands, industries are encouraged to use "benchmarking covenants" to become the most energy-efficient industries worldwide. In return for industry cooperation and progress, the government agrees not to impose overly stringent policies on firms. This approach represents an interesting application of a nodality tool (benchmarking) that is rendered more effective by promises to refrain from applying another policy instrument (in this case, a regulatory instrument).[36] In Singapore, the Building and Construction Authority has had a building energy use benchmarking program in place since 2008. Consequently, in conjunction with energy audit subsidies, overall energy use in the nation's buildings improved by 11 percent by the end of 2017.[37]

According to the IPCC, the factors critical to the success of benchmarking programs include "a strong institutional framework, a robust and independent monitoring and evaluation system, credible mechanisms for dealing with non-compliance, capacity building and, accompanying measures such as free or subsidized energy audits."[38]

Numerous *environmental disclosure programs* have also developed over the last decade to document corporate environmental performance, sustainability, and carbon emissions and to allow interested stakeholders to evaluate environmental governance. Perhaps the best known of these tools is the ISO 14000 family of standards, which originated in the EU in support of the EU's Eco-Management and Audit Scheme (EMAS) and which have subsequently diffused to many other regions, including East Asia (China, Taiwan, South Korea, and Japan), which is the world's fastest-growing ISO 14000 market.

One of the criticisms of the ISO 14000 series is that the standards are not prescriptive enough to induce a substantive change in performance.[39] Consequently, a number of more prescriptive disclosure programs have arisen.

The Carbon Disclosure Project (CDP), which was founded in the United Kingdom, is one of the most influential of these systems. Supported by an international, not-for-profit organization that acts on behalf of 767 institutional investors, it helps to evaluate the climate change–related risk in their investment portfolios. After an initial phase with spotty reporting, the CDP has become more consistent and transparent, particularly in the treatment of different types of emissions.[40] Scope 1 emissions account for direct emissions from a facility, Scope 2 emissions cover electricity use, and Scope 3 emissions track emissions from the entire product life cycle, supply-chain, and distribution systems.[41] Disclosure programs such as the CDP promise to be influential in motivating corporate energy efficiency because published corporate performance indicators are increasingly evaluated by direct investors and customers, not to mention the numerous green investment funds that are burgeoning worldwide.[42]

In keeping with the theme that better information makes markets operate more efficiently, many countries sponsor *energy audit programs*, which integrate energy management systems that include maintenance checklists, measurement systems, performance indicators, and progress reporting.[43] Evaluations of such audit programs in Germany, the United States, and Japan indicate that significant energy savings have resulted, while costs of implementation are modest.[44]

Energy audits are a staple in nations that host energy-intensive industry—including Australia, China, Japan, South Korea, and the United States—because energy efficiency often equates with enhanced competitiveness. Through energy audits, industries can identify energy efficiency opportunities, which often lead to subsequent participation in financial assistance programs. Singapore's approach to energy audits reflects the importance of pairing nodality tools (energy audits) with other tools to incentivize uptake. The National Environmental Agency (NEA) in Singapore offers subsidies to firms to explore energy savings; it provides support for energy audits and licensing accreditation programs to enable energy managers and energy service companies (ESCOs) to undertake audits. The nation even has award schemes for firms that make substantial progress.[45]

For many firms, investments in energy efficiency can be highly profitable. In Germany, energy audits delivered via ESCO support have helped small and medium-sized enterprises (SMEs) bridge the competency gap so that they can exploit technology for enhanced efficiency.[46] This strategy was found to be quite cost-effective in boosting SME competitiveness.[47] This was also the case with the energy audit program implemented by the Energy Conservation Center of Japan.[48]

In the United States, the Department of Energy's energy audit program includes a unique collaborative structure. It is implemented through twenty-four university industrial assessment centers, with faculty and students performing energy audits on nearby small and medium-sized industrial sites.[49] Research suggests that this model could be even more effective if it were able to leverage relationships with financial institutions and engineering firms, thereby enhancing workforce prospects for university students in engineering and business.[50]

In sum, nodality instruments such as benchmarking and labeling spur energy efficiency investments when implemented on their own, but they can be even more powerful when coupled with other policy instruments.

Common Authority Instruments

Authority instruments—laws, regulations, and mandated standards—are coercive tools that compel stakeholders to alter behavior or face negative sanctions. In policy parlance, these are known as "sticks" because failure to comply with authority tools can mean government censure, loss of operating permits, fines, and even prison time.[51] Although authority instruments can be highly effective when rapid compliance is required or the problem being curtailed is of a heinous nature, these tools are also rigid (thereby leaving little leeway for firms to adapt in a cost-effective manner)[52] and can be costly to enforce because authority tools require overseers to monitor compliance.[53]

However, there is an emerging perspective on regulatory intervention that, albeit rooted in strategic management theory, runs counter to free-market ideological convention. Harvard strategic management professor Michael Porter and colleague Claas Van Der Linde published a perspective on regulations that, in the era of free trade in 1995, was surprising. Cruelly ignoring Van Der Linde's co-authorship, this became known as Porter's hypothesis. Porter's hypothesis suggests that well-crafted environmental regulations can trigger innovation, resulting in enhanced environmental governance *and* enhanced corporate profitability. The premise is that stringent regulations, if applied evenly to all competitors, not only ensure a higher aggregate level of environmental performance but also reward firms that innovate and move up the productivity curve.

Data support the premise that corporate environmental research and development (R&D) is enhanced when stringent environmental policy is introduced.[54] An increase in patents following regulations of certain environmental pollutants has been documented in the United States, Germany, and Japan.[55] Notably, though, researchers have also found that flexible regulations such as performance-based standards that allow customized solutions have a more positive impact on environmental and business performance, compared with technology mandates.[56]

The Porter hypothesis suggests that, in the energy sector, mandating companies to adopt renewable energy technologies could improve environmental governance and also increase profits for power generation companies by promoting productivity-enhancing innovation. Similarly, stricter building codes and appliance standards could also seed innovation.

Motives aside, there are a number of prominent regulatory tools in the energy sector that can help seed a transition. Many of these authority tools are supply-side tools: regulations designed to influence the behavior of energy generation firms. These *supply-side regulations* can be packaged in different ways; they can use mandatory renewable energy targets, renewable energy quotas (including renewable portfolio standards), emission caps, performance standards, and building codes. Like nodality tools, authority tools are rarely applied in isolation; rather, they are part of a policy portfolio that includes other supporting instruments.

Renewable energy targets have been influential in promoting clean-tech options because they serve a powerful signaling role that influences energy generation investment patterns. Australia established a goal of 23.5 percent renewable electricity generation by 2020,[57] and the European Union is aiming for 50 percent renewable electricity by 2030.[58] India's target under the Paris Agreement is 40 percent of total electricity capacity from non–fossil fuel sources by 2030.[59] Recently, China's thirteenth five-year plan for the power sector calls for the percentage of non–fossil fuel generating capacity to increase to 35 percent by 2020.[60] Despite truculence at the federal level, as of 2018, thirty U.S. states had renewable energy targets, and eight more had voluntary targets.[61]

The signaling role of supply-side targets is better illustrated through the Intended Nationally Determined Contributions (CO_2 emission targets) to the Paris Agreement of four major carbon-emitting nations, as summarized in table 6.2. Today these nations account for 62 percent of global emissions. The variance in their CO_2 targets makes it difficult to compare and contrast their relative aspirations, but some are clearly more ambitious than others.

To meet their declared targets, many nations establish *mandatory renewable energy quotas* for energy providers. One of the most common manifestations is the *renewable portfolio standard* (RPS). An RPS is a mandate given to energy providers that a certain percentage of total power generation must be provided through renewable energy by a given date. If power generators fail to comply, they are usually fined. In the United States, thirty of the fifty states had RPS legislation in place in 2018.[62] RPSs have also been used at the national level in the United Kingdom and Japan, among other nations, but have been criticized as being overly rigid and therefore less effective than other market-based approaches. Moreover, as was the case in Japan, lukewarm standards can actually stymie progress by giving generators an excuse for inaction.[63] In the transport sector an equivalent policy is a *blending mandate*, which compels petrol retailers to blend biofuel with petrol. For example, Belgium passed a royal decree in the summer of 2016 that set a binding target of 8.5 percent biofuel starting from 2020.[64]

TABLE 6.2 Energy consumption, CO_2 emissions, and emission targets of four major carbon-emitting nations

	Energy use (quadrillion Btu)	CO_2 emissions (MMmt)*	Population (millions)	Carbon footprint (mtCO$_2^\dagger$/ capita)	CO_2 emission reduction target (compared to 2005 level)	GDP (billion 2010 $)
United States	97.0	5,146	325	15.8	26–28% by 2025 vs. 2005[‡]	16,512
OECD-EU	80.7	3,930	573	6.9	40% by 2030 vs. 1990[§]	20,321
China	136.3	10,010	1,376	7.3	60–65% per unit of GDP by 2030 vs. 2005[∣]	9,510
India	29.8	2,108	1,311	1.6	33–35% by 2030 vs. 2005**	2,354
World	580.7	34,095	7,336	4.7		76,796

Source: Marilyn A. Brown, Shan Zhou, and Majid Ahmadi, "Evolving Smart-Grid Policies: An International Review," *Wiley Interdisciplinary Reviews (WIREs): Energy and Environment*, 2018.

*MMmt =

[†]mtCO$_2$ = metric tonnes of CO_2.

[‡] "U.S. Cover Note INDC and Accompanying Information.pdf (Rep.)," March 31, 2015, http://www4.unfccc. int/Submissions/INDC/Published%20Documents/United%20States%20of%20America/1/U.S.%20Cover%20 Note%20INDC%20and%20Accompanying%20Information.pdf.

[§] "Intended Nationally Determined Contribution of the EU and Its Member States (Rep.)," March 6, 2015, http://www4.unfccc.int/Submissions/INDC/Published%20Documents/Latvia/1/LV-03-06-EU%20INDC.pdf.

[∣] "Department of Climate Change, National Development and Reform Commission of China (Rep.)." June 30, 2015, http://www4.unfccc.int/Submissions/INDC/Published%20Documents/China/1/China's%20INDC%20 -%20on%2030%20June%202015.pdf.

** "India's Intended Nationally Determined Contribution: Working Towards Climate Justice (Rep.)," http:// www4.unfccc.int/Submissions/INDC/Published%20Documents/India/1/INDIA%20INDC%20TO%20UNFCCC .pdf.

In order to provide more flexibility to power generators, *emission caps* have emerged as attractive policy options. Under an emission cap system, a regulatory authority establishes legislation that forces power providers to keep emissions under a prescribed limit. Any power supplier that exceeds the cap is either fined or, in extreme cases, is subject to plant closure. The allure of a cap system is that it directly addresses the objective (lower emissions) but gives power providers flexibility over how to achieve the objective. A cap has been embraced by a number of nations because it can be combined with an emissions trading regime to allow power providers that are most effective in reducing GHG emissions to sell some of their savings to less effective power providers, thereby providing a financial reward to the transitioning firm. This is known as *cap and trade*.

Cap-and-trade strategies have been used at most policy levels. Globally, a cap-and-trade system exists in the form of the Kyoto Protocol Clean Development Mechanism and Joint Implementation programs.[65] Regionally, the EU Emissions Trading Scheme (EU ETS) is easily the most comprehensive in that it operates in thirty-one nations and covers more than 45 percent of the GHG emissions in the member states.[66] The allure of a cap-and-trade system at the transnational level is that it allows nations at different stages of economic development to participate in transitionary efforts by hosting clean energy projects that can be established at lower cost in low-cost nations. At the national and subnational levels, at the end of 2017, there were at least nineteen systems in place around the world, governing more than 7 billion tons of GHG emissions. These included national-level programs in South Korea, Switzerland, and New Zealand; provincial- and state-level programs in China, Canada, the United States, and Japan; and municipal programs in China and Japan.[67]

Although cap-and-trade programs can be effective in curtailing supply-side emissions (e.g., from power plants), other approaches have proven to be more successful at incentivizing technological innovation on the demand side. Many countries have implemented *appliance and equipment standards* to eliminate energy-guzzling appliances and equipment from the marketplace. These minimum energy performance standards (MEPS) typically target home appliances, consumer electronics, water heaters, space heaters, cooling equipment, and industrial motors.

The technological journey of the household refrigerator illustrates the potential for improving efficiency by implementing performance standards paired with R&D support. Between 1977 and 1982, the U.S. Department of Energy (DOE)'s research programs sired improved refrigerator compressors, motors, insulation, and controls. California then set state appliance standards to encourage mass market adoption, and shortly thereafter the DOE promulgated national standards. Over the years, the regulations were ratcheted up so that by 2010 the energy consumption of a new home refrigerator had declined by 70 percent.[68] On the heels of programs that have significantly improved gas furnaces and central air conditioners[69] and essentially eliminated incandescent bulbs, DOE standards are now transforming the markets for unit air conditioning and heat pump systems. As the DOE website boasts, "The products regulated by the program represent about 90 percent of home energy use, 60 percent of commercial building energy use, and 30 percent of industrial energy use. Standards saved American consumers $63 billion on their utility bills in 2015, and cumulatively, have helped the United States avoid 2.6 billion tons of carbon dioxide emissions."[70]

Appliance standards have gained prevalence among many national, state, and provincial governments. At about the same time that the United States was creating the bulk of its appliance standards programs, the EU's ecodesign scheme and Australia's MEPS program were also under way, leading to the diffusion of appliance standards to many other countries through market pressures. Not to be outdone, Japan subsequently implemented appliance standards that were based on a careful monitoring of technological trajectories and designed to reflect emerging best practice; these standards resulted in a superior level of adeptness and innovation.[71]

Conversely, industrial technology standards seem to have had less success, largely because of the varied and customized nature of industrial equipment. It is hard to standardize things that are customized. One key exception is industrial motors, which have been the subject of appliance standards in many countries. The IEA recommends adopting MEPS for all motors.[72] In the United States, the Energy Independence and Security Act of 2007 (EISA) upgraded standards on all new motors sold in the United States from requirements previously laid out in 1992. India has a similar policy strategy that promotes the replacement of conventional motors with energy-efficient motors in industrial firms.[73]

In a nutshell, *product standard setting programs* can engender significant energy savings.[74] In addition to providing immediate savings through technology upgrades, they also (a) directly affect investment decisions; (b) assist the diffusion and market penetration of energy efficiency measures; and (c) initiate cross-national spillover effects when other countries follow suit. Critics claim that command-and-control approaches such as enforced standards impede innovation; but others counter that standards incentivize innovation if sufficient flexibility is afforded manufacturers to develop new technologies in order to meet the minimum efficiency requirements. To synthesize these perspectives, one can suggest that standards should be viewed as shapers of markets rather than hurdles or triggers of innovation.[75]

Mandating and enforcing stringent *energy codes for buildings* can also reduce energy consumption and save building owners and occupants money.[76] Building energy codes have been found to be cost-effective in many jurisdictions[77] and are often seen as a component of better design, a prevalent perspective evident throughout Europe. In 2002, the European Parliament and Council promulgated Directive 2002/91/EC on the Energy Performance of Buildings, which was motivated by the EU's commitment to the Kyoto Protocol to reduce its emissions by 5.2 percent below 1990 levels by 2010. The Directive required each member state to develop measures that can be applied at a regional or national level to meet the minimum standards on energy efficiency for new buildings and existing buildings that go through major renovations. The performance of buildings had to be documented through energy performance certificates that display information to the public about the energy performance of a building and how it compares to legal standards. Each member state was required to set up efficiency standards and measure the performance of buildings in terms of the building shell, heating and cooling installations, hot water supply, ventilation, lighting systems, building orientation, passive solar systems, and other factors. Combined heat and power and district heating and cooling were also part of the desired design features. As table 6.3 illustrates, the subtleties of what is covered in building energy codes tend to reflect national context and policy priorities.

Regional initiatives, such as the EU directives, promote national policy innovation. For example, the German Energy Conservation Act (EnEV) requires periodic code updates to respond to technical developments.

TABLE 6.3. Key features of building energy codes

Country	Level of implementation	Scope covered							Regulate efficiency or emissions	Nearly Zero-Energy Consumption by 2,0	Renewable Energy Included	Certification Program Included
		Space heating	Space cooling	Ventilation	Water heating	Lighting	Building Envelope	Design, position and orientation				
United States	State by state	✓	✓	✓	✓	✓	✓	✓	Energy efficiency	No	No	No
Germany	National	✓	×	✓	✓	✓	✓	✓	Energy efficiency	Yes	Yes	Yes
United Kingdom	National	✓	×	✓	✓	✓	✓	✓	CO_2 emissions	Yes	Yes	Yes
China	National (some rural areas are exempted from inspection)	✓	✓	✓	×	×	✓	✓	Energy efficiency	No	No	No

Source: Xiaojing Sun et al., "Mandating Better Buildings: A Global Review of Building Codes and Prospects for Improvement in the United States," *Wiley Interdisciplinary Reviews (WIREs): Energy and Environment* 5, no. 2 (2016): 188–215, doi:10.1002/wene.168.

The tenet that governs the EnEV is that mandated construction and retrofit measures must be able to recover costs over their useful lifetimes.[78] Innovative approaches also exist for (1) expanding code adoption, (2) enforcing code compliance, and (3) promoting performance above code.

Building code innovations are also being developed to foster more efficient construction practices. In May 2010, the EU adopted a directive stipulating that by the end of 2020 all EU member states must ensure that newly constructed buildings consume "nearly zero energy" or are climate-neutral (both energy efficiency and the use of renewable energy count toward meeting this goal).[79] In June 2018, a directive (2018/844/EU) amending the 2010 Energy Performance of Buildings Directive was published. The revision introduces amendments to accelerate the cost-effective refit of existing buildings, with the aim of achieving complete decarbonization of EU building stock by 2050. The plan also introduces provisions to enhance smart technologies. Member states will have twenty months (by March 10, 2020) to embed its provisions into national law.[80]

Despite evidence of political will, the enforcement of building codes has been found to be inadequate in the United States and Europe.[81] Implementing random audits was shown to strengthen enforcement in Germany.[82] An inadequately trained workforce was found to be the cause of enforcement shortfalls in many developing countries.[83] A promising approach to tackling the compliance shortfall is to create a liability framework where parties subject to building codes have clear responsibilities and are held accountable through third-party inspections and penalties for noncompliance.[84] China uses such a system to ensure compliance with standards for energy efficiency in new buildings.[85] In sum, for both of these regulatory approaches— product standards and building codes—developed and developing countries alike increasingly recognize the need for stronger enforcement.[86]

Common Treasure Instruments

If regulations are considered the sticks within the policy instrument tool box, *treasure instruments* can be considered the carrots because they engender desired change through financial incentives or disincentives.[87] In the energy sector, treasure instruments prevail both on the supply side—through instruments such as *renewable energy production subsidies, carbon taxes, investment tax credits,* and *research subsidies*—and on the demand side— through instruments such as *consumption taxes,* rebates for energy-efficient goods, and *tax credits* for investing in more efficient technologies.

The most prevalent form of supply-side subsidy is the *feed-in tariff (FIT).* FITs guarantee grid access (there is an authority element to this instrument) over an extended period (typically fifteen to twenty years) with prices based on the cost of generation plus a rate of return that is intended to draw in sufficient firms to meet capacity development targets (this is the treasure element). FIT payments are usually administered by the utility and can be compensated for through an additional charge for electricity imposed on national or regional customers.

At least eighty-four countries had FITs at the close of 2017.[88] Germany's FIT was one of the first and is still one of the strongest. Starting with 4.8 percent renewables in 1998, Germany's share of renewable electricity grew to 28 percent in 2014.[89] In 2012, after the Fukushima disaster, the Japanese government finally turned it sights to seriously incentivizing renewable energy development and launched a series of FITs that were and still are some of the highest in the world. Table 6.4 provides a summary of how Japan constructed its FIT program. In 2015, the government collected over US$130 million in premiums from energy consumers to support this program. To convey how strong these premiums are, in December 2018, 10 Japanese yen (JPY) equated to US$0.09.

Whereas FITs incentivize renewable energy providers, *carbon taxes* disincentivize fossil fuel energy providers. Carbon taxes can be percentages or fixed dollar amounts tied to fossil fuel use (e.g., cents per liter for petrol or cents per kilowatt hour for coal-fired power plants). Either way, these taxes help to level the competitive playing field by ensuring that the social and environmental costs of fossil fuels are somewhat reflected in the prices.

Which approach a nation adopts and how the tool is structured depends largely on the political context. For example, since the inception of its modern-day wind power development program, Germany has relied on a feed-in tariff to drive development.[90] However, within the EU there is a trend toward competitive auctions of renewable energy, and all member

TABLE 6.4 Japan's feed-in tariff program

		Purchase prices (JPY*/kWh) (tax excluded)						Purchase period
		FY2012	FY2013	FY2014	FY2015 Apr.–Jun.	FY2015 Jul.–Mar.	FY2016	
Solar	Less than 10 kW	42	38	37	33		31	10 years
	When output control systems are required				35		33	
	Less than 10 kW (+ energy storage system)	34	31	30	27		25	
	When output control systems are required				29		27	
	10 kW or more	40	36	32	29	27	24	20 years
Wind	Onshore Less than 20 kW	55	55	55	55		55	20 years
	Onshore 20 kW or more	22	22	22	22		22	
	Offshore			36	36		36	
Geothermal	Less than 15,000 kW	40	40	40	40		40	15 years
	15,000 kW or more	26	26	26	26		26	

(continued)

TABLE 6.4 (*continued*)

			Purchase prices (JPY*/kWh) (tax excluded)						Purchase period
			FY2012	FY2013	FY2014	FY2015 Apr.–Jun.	FY2015 Jul.–Mar.	FY2016	
Hydro	Fully new facilities	Less than 200 kW	34	34	34	34	34	34	20 years
		200–1,000 kW	29	29	29	29	29	29	
		1,000–30,000 kW	24	24	24	24	24	24	
	Utilize existing headrace	Less than 200 kW			25	25	25	25	
		200–1,000 kW			21	21	21	21	
		1,000–30,000 kW			14	14	14	14	
Biomass	Wood (general), agricultural residues		24	24	24	24	24	24	20 years
	Forest residues	Less than 2,000 kW				40	40	40	
		2,000 kW or more	32	32	32	32	32	32	
	Wood waste from buildings		13	13	13	13	13	13	
	Municipal waste		17	17	17	17	17	17	
	Biogas		39	39	39	39	39	39	

Source: https://www.iea.org/policiesandmeasures/pams/japan/name-30660-en.php.

*JPY = Japanese yen.

nations are compelled to adopt such policies domestically. Consequently, the German Renewable Energy Act of 2017 altered national policy from a FIT system to a tendering system to inject greater price competition into the renewable energy sector.[91] As reported earlier, Japan has elected to go with a FIT system, but because the government is fiscally challenged, it has elected to fund the program through a surcharge on electricity consumed. Conversely, other nations fund FITs through fiscal expenditure. For carbon taxes, some nations tax fossil fuel emissions and use the funds to bolster tax revenues for use in other areas, while other nations redistribute the carbon taxes to renewable energy suppliers to accelerate the transition.

A compromise treasure instrument is an *investment tax credit*, which provides tax credits to inefficient energy providers for upgrading technology and provides renewable energy providers with tax credits for building renewable energy facilities. India, for example, has given tax breaks to renewable energy providers of all types since 2014.[92] On the one hand, investment tax credits are attractive to policymakers because they do not cannibalize on existing government funds, and they are attractive to project developers because in profitable years surplus cash can be sheltered from taxation through such investments. On the other hand, investment tax credits have fallen out of favor in some nations because they incentivize building facilities rather than actually producing energy. In response to this concern, some nations—notably the United States— have adopted a *production tax credit* system, instead of an investment tax credit. In this way, project developers have to create solutions before they are rewarded.

A final supply-side treasure instrument category worth noting includes *research and development subsidies* and *research tax credits*. Both of these instruments are important additions to the policymaker's tool kit because R&D subsidies or tax credits encourage firms to invest in technological innovation. As opposed to the instruments outlined earlier, these incentives shift the focus from short-term results to medium- or long-term results and help build competitive industries. Tax credits and incentives have been successfully employed in Denmark, the United States, Germany, China, and India, and world-class energy firms have been nurtured through these instruments.

Turning to the demand side, a number of highly effective treasure instruments are commonly found in the energy world. The first category, *consumption taxes*, represents instruments that virtually everyone is familiar with. In all developed nations, gasoline taxes exist, albeit at different levels of rigor. For example, in petroleum-rich Canada, a federal gasoline tax of US$0.08 per liter[93] combines with provincial excise taxes that average US$0.11 per liter. In Australia, the excise tax for petrol averaged about US$0.31 per liter in 2017.[94] Meanwhile, in the United Kingdom, the taxes on petrol in 2018 amount to a whopping US$0.78 per liter.[95] In theory, consumption taxes can have a dual effect on market behavior. Taxes raise the cost of consuming energy and encourage less use. Additionally, the tax revenues raised can be channeled into subsidies to encourage the development of better energy technologies. Unfortunately, in practice this rarely happens. What typically happens is that governments apply these petrol taxes and become dependent on the revenue flows to help fund other areas of governance. Ironically, these revenue flows then become shackles that inhibit transition because governments find it too difficult to wean themselves off the riches that the taxes provide.

Recently, a number of governments have been providing *rebates* to consumers to encourage them to switch to lower-carbon or more energy-efficient technologies. These rebates are targeted widely from specific sectors to individual consumers. For example, the Ministry of Environmental Protection (MoEP) in Israel has set aside approximately US$6.5 million to encourage bus companies to upgrade to electric buses.[96] The Australian government launched a US$4 million program in 2016 entitled Solar Communities that was designed to provide funding for community organizations that wish to install rooftop solar PV, solar hot water, or solar-connected battery systems.[97] Since 2015, the Indian state of Tamil Nadu has provided a subsidy to individuals or groups interested in connecting rooftop solar PV systems of up to 1 kW for residential purposes. The program also permits consumers to export excess energy into the grid as an offset against energy drawn from the grid.[98] As an example of technological switching subsidies, the Mexican government ran an Electric Power Savings Trust Fund (FIDE) from 2002 to 2012 that helped finance an upgrade of refrigerators and air conditioners, as well as other energy-saving measures. In the first phase alone, the program led to the replacement of 130,000 refrigerators and 623,000 air conditioning units.[99]

The downside to rebates is that—get ready for the shocker—they cost money. More specifically, these types of programs can draw down fiscal budgets. Consequently, some governments prefer to incentivize technological upgrading by offering *technology upgrade tax credits* instead. For example, Singapore's National Environmental Agency provides grants to industry for investing in technological upgrades and for engaging energy consultants to aid in energy efficiency upgrades.[100] Under its Energy Efficiency Fund (E2F), Singapore also provides grants to new companies to encourage "investors of new industrial facilities or major expansions in Singapore to integrate resource efficiency improvements into manufacturing facility development plans early in the design stage."[101]

Common Organizational Instruments

Within the NATO framework, organization instruments are essentially the boots-on-the-ground initiatives that governments support through agencies, advisory boards, public or industry one-stop energy shops, public think tanks, public universities, and private-public partnerships. The allure of direct government provision of energy services, education and consulting services, and research is intuitively obvious: it affords government more control over the entire policy process, from policy design to implementation.[102] The downside is that government implementation does not always achieve optimal levels of effectiveness.

In the energy sector there have been some particularly adept uses of organization instruments and some valuable lessons learned. First and foremost is the structuring of electricity provision infrastructure. Because energy is such a strategically important commodity, many nations, ranging from Taiwan and China to Kazakhstan and Canada, nationalize energy provision by establishing public utilities. However, a common criticism of this approach is that it leads to inflated energy costs because of the inefficiencies within the public utility and the lack of competition. Consequently, the model that seems to have gained the most traction as nations begin to explore ways to integrate the cheapest possible renewable energy contributions is a *public grid operator* that is mandated to accept electricity flows from *private energy producers*. This model strikes an effective balance in that competition at the generation phase is encouraged while grid-level economies of scale and security can be effectively controlled by national grid operators.

Regardless of how electricity provision is structured, the strategic importance of electricity in supporting economic development and human welfare motivates nations to establish *regulatory authorities* to oversee the process. The role of electricity regulator differs by nation, but common tasks include administration of taxes, regulation of energy prices, enforcement of operational regulations, and market coordination to ensure that there is sufficient supply to meet demand. As more renewable energy capacity is added to the energy mix, these regulatory authorities serve an important role in ensuring that the process is managed so that communities are not adversely affected (e.g., by overseeing wind power development codes in Denmark), private utilities are not erecting artificial barriers to stymie renewable energy contributions to the grid (as has been the case in Japan in the past),[103] and the system is working for all (as opposed to the situation in China, where some new wind power installations were left unconnected to the grid).[104]

When it comes to the design of policy in the energy sector, a recent development is the emergence of *specialized government departments* that are charged with the task of investigating and analyzing energy policies and innovating new policy solutions. For example, the Singaporean Civil Service College has a specialized unit that investigates enhanced policy through behavioral nudges. Singapore Land Transport Authority (LTA) also has a team of behavioral economists who work on similar strategies in regard to influencing use of public transportation. As a result, between 2012 and 2014, the LTA launched a "Travel Smart Rewards" scheme designed to incentivize the public to shift travel to off-peak hours in order to ease peak travel congestion.[105] The scheme centered around an online game that rewarded customers for off-peak trips, with cash prizes as high as US$1,100. The program resulted in a 10 percent shift from peak to off-peak travel among participants.[106] Similarly, the U.K. government enjoys an ownership stake in a "social purpose company" called Behavioral Insights. This team employs nudging techniques to try to affect behavioral change. For example, in one of its experiments, smart thermostats were installed in 2,248 homes to evaluate the impact on energy usage of real-time information being fed to the consumer. The study found that the smart meter led to savings of 6 to 7 percent of annual heating fuel use.[107] The allure of these types of governmental departments is that insights could lead to substantial savings in policy implementation.

Another effective organizational instrument in the renewable energy sector is the creation of *government R&D hubs*. Examples include the National Renewable Energy Laboratory in the United States; the government-supported Risø DTU at the Technical University of Denmark, which was instrumental in supporting technological capacity building in wind power during the inception period of the Danish wind power program;[108] and the numerous publicly funded R&D iniatitives established in the EU, with support from Horizon 2020 funding.[109]

The Risø DTU can also be considered a form of a new government entity—*government consulting services and information centers*—that have cropped up recently to serve as a one-stop shop for helping educate stakeholders on how to play an active role in facilitating the energy transition. In addition to the R&D that is done at Risø DTU, the organization also offers educational events and provides access to experts so that investors who are interested in renewable energy can learn more about the opportunities available. While Risø DTU serves as an example of supporting investment in renewable energy, there are far more of these entities emerging in communities to provide consulting services on how to improve energy efficiency for businesses and residential energy consumers. Regulatory initiatives to promote the efficient consumption of energy have multiplied across the globe over the last several decades, and information centers are proving to be effective in disseminating much-needed information.[110]

For example, Efficiency Vermont in the United States was created in 1999 by the Vermont Legislature and the Vermont Public Utility Commission to help reduce energy costs for residents of Vermont. It offers energy assessments, access to financing, advice on renovation and construction, and project support and holds educational events to achieve its mission. According to its website, the organization has catalyzed 14.5 MWh of lifetime electricity savings since 2000.[111]

As another illustration, Oberösterreich Energiesparverband (ESV), the energy agency in the region of Upper Austria (near Linz), boasts a central information hub that provides guidance and support to households, businesses, and municipalities that wish to adopt new energy systems or improve energy efficiency. The center's staff provides expert advice through over 10,000 face-to-face sessions each year, either at ESV offices or through home visits. The first 1.0 to 1.5 hours are free. It also has a hotline that receives on average 80 phone calls each day, and its website (www.esv.or.at) is frequented by 500 residential consumers of energy. On the business side, the center advises, on average, 200 businesses each year. For businesses, this is not a free service; 75 percent is covered through regional and national programs, with the remainder paid by the business.[112] As will be detailed in the next chapter, these advisory services are proving to be very useful for bridging knowledge gaps that would otherwise impede action by residential or even commercial energy consumers.

A final organizational instrument worth noting is *public-private partnerships (PPPs)*. The energy transition will require trillions of dollars in investment and will significantly reshape energy networks. Accordingly, PPP arrangements are seen as holding promise for a number of prominent reasons. First, most governments simply do not have the funds to direct such a transition on their own. Second, governments do not have the expertise to drive a transition that will ultimately rely on numerous technologies cobbled together in a manner to deliver cheap and effective energy services. Third, when markets are disrupted, there will be extreme opposition if stakeholders are not given a voice in the process. PPPs provide both a voice and a reward for collaborating.

As a result of the benefits of PPPs, this organizational instrument is found in new renewable energy projects, R&D hubs, and information centers. In developing nations, renewable energy PPPs are often driven by international agencies such as the World Bank's International Finance Corporation, which has improved electricity access for 10 million people through work with private energy providers, who have contributed over US$1.3 billion to the construction of 1.3 GW of new renewable generation capacity.[113] Yet PPPs are not only a developing nation strategy. For example, Japan makes extensive use of PPPs for driving smart city initiatives.

In its approach to energy transformation, consortiums are created to match local government needs with companies that can manage the technological transformation. Projects delivered through this model have ranged from wide-scale introduction of renewable energy and electric vehicles in Yokohama to a pilot project that saw the installation of 700 home energy management systems in Keihanna (Kansai region).[114]

THE LOGIC OF POLICY INSTRUMENT PORTFOLIOS

Policy instruments are rarely designed and implemented as stand-alone tools. Individually, policy instruments tend to lack elements that are needed for the policy to be effective. Take, for instance, the Perform Achieve Trade (PAT) scheme—an innovative, market-based trading scheme included in India's National Action Plan on Climate Change that was launched in 2008.[115] The scheme aimed to improve energy efficiency in energy-intensive industries by creating a trading platform for energy efficiency certificates. But to make this treasure instrument work effectively, stakeholders needed to know how it worked, so nodality instruments were necessary. Moreover, without some form of coercion, there was no incentive for firms to embrace the program in the first place. Therefore, regulations were needed to establish targets upon which the certificate system could be founded. Consequently, an amendment to the Energy Conservation Act (ECA) was made in 2010 to provide a legal mandate to PAT. Participation in the scheme was made mandatory for Designated Consumers under the ECA. The PAT scheme is being implemented in three phases; the first phase ran from 2012 to 2015 and covered 478 facilities from eight energy-intensive sectors (aluminum, cement, chor-alkali, fertilizer, iron and steel, pulp and paper, textiles, and thermal power plants). These facilities account for roughly 60 percent of India's total primary energy consumption.[116]A facility's baseline is determined by its historic specific energy consumption between 2007 and 2010. The average reduction target was 4.8 percent in 2015. Facilities making greater reductions than their targets received energy-saving certificates that can be traded with facilities that are having trouble meeting their targets, or banked for future use. To summarize, a profit motive incentivizes investments in energy efficiency, but this treasure tool would not have been viable without regulatory support, nodality instruments to educate the participants, and an organization structure to supervise program implementation and manage the certificate verification process. This is commonly the case with policy instruments: they work best in portfolios.

So how does one choose the appropriate policy instrument mix? In many ways policymaking is as much a craft as it is a science. Certainly there is a base of empirical evidence to support the effectiveness of bundling different policy instruments, but at the end of the day, applying policy depends on an accurate understanding of the problem, the behavior of stakeholders, underlying motives, and a host of social, technological, economic, environmental, and political factors that evolve over time and largely frame the eventual effectiveness of any given policy mix.

With this knowledge in mind, we now turn to a chapter that highlights the importance of understanding stakeholders, motives, and habitual patterns in order to develop strategies to counter complacency and, dare we say it, consumer apathy. The reader will discover how understanding stakeholders and their contextual environments allows us to select the tools from this chapter that are most likely to affect change.

7 | Consumers as Agents of Change

When it comes to consumer behavior, there is good news and bad news. The good news is that consumers are slowly adopting more sustainable consumption practices. This change is largely driven by the ineluctable link between energy usage and climate change.[1] Moreover, because most governments view progress in energy efficiency as desirable, many public authorities are incentivizing more productive energy practices.[2] In short, the fact that people seem to be increasingly aware of the need to rein in consumption and that governments are sympathetic to this goal is something that should leave us with knees wobbling in pleasure.

Now for the bad news: we live in a world of energy addicts. Indeed, dear reader, you are likely one yourself. Many homes in the industrialized world boast an inventory of "essential" energy-sucking equipment, including air conditioners, microwave ovens, bread makers, dishwashers, clothes dryers, tropical fish tanks, massage chairs, ice makers, home theaters, electric can openers, electric blankets, electric clocks, and of course, the hallowed "beer fridge." Worldwide, householders own approximately 1 billion personal automobiles—giant boxes of steel, wire, and plastic that require 95 percent of the gasoline consumed just to push these heavy boxes around.[3] The energy requirements for the production, transportation, and management of food, household goods, and services amount to about half of total household energy use in Europe.[4]

In short, we are energy addicts who are just now starting to realize that we might have a problem. We still have a long road ahead of us. Despite much ballyhooing, there exists a clear misalignment between our aspirations and our behaviors as consumers. Even environmentalists find it hard to live up to the change that is needed. Overseas trips to speaking engagements, domestic hops between cities, leisure plans—these all consume quantities of energy that we can ill afford to consume over a sustained period. We might be aware of this disconnect, but at the end of the day, we are all still addicts—next week, we will reform. Always next week.

However, it is promising that choices made regarding what goods and services are purchased (and in what quantities) and how they are utilized in everyday life can make a huge difference in reducing our collective energy footprint.[5] The question is, how can we bypass the denial and opposition to change? To answer this question, we first need to understand the plight of the addict. What drives consumption behavior? How did we get here, and how can we reverse direction?

In this chapter, we first illustrate the various roles that we play as consumers. We then explore a new type of consumption behavior—that of the prosumer—before drawing from innovation studies, science and technology studies, sociology, and psychology to clarify four ways in which the consumption process can be modified. Finally, to better define the challenges ahead, we turn to an investigation of barriers that inhibit change, including apathy and ambivalence, ignorance and misunderstanding, culture and customs, and active (and concerted) resistance. In concluding, we demonstrate how change is possible by putting forth four case studies from the Netherlands, United Kingdom, United States (Vermont), and Brazil where behavioral modification programs have had some success.

USERS: CONSUMERS, AND PROSUMERS

How well do you think you understand the role of consumers in influencing market change? This section explores how consumer behavior shapes energy sector developments. To date, the difference the consumer has made has largely been negative, but this need not be irreversible. If consumers have sweeping powers to adversely impact energy market behavior, they also have the ability to reverse their habits and, in doing so, invoke positive change.

The term *consumers* ordinarily refers to direct consumers—individuals, households, firms, or industrial buyers who are using energy services. Using this definition, the goal is therefore to simply alter the consumption process at the end-user level. However, recent work has broadened the perspective to also include indirect consumers or user intermediaries as influential members of the consumption network.[6] Simply put, other stakeholders also influence consumption behavior. For example, drivers or passengers are not the only consumers who have a stake in influencing transportation sector behavior. Other influential stakeholders include investors who encourage growth of certain firms, engineers who design highway systems, mechanics, insurance executives who alter consumption patterns through the rates that they set, and even traffic police who influence driving practices.

Schot and colleagues identify five types of users when conceptualized from this broader perspective:[7]

- *User-producers* (e.g., firms) create new technical and organizational solutions.
- *User-intermediaries* (e.g., marketing firms, petrol stations) shape the needs and desires of users as well as supporting product features, providing infrastructure, and influencing regulatory frameworks.
- *User-citizens* engage in politics and lobbying for a particular niche or cause.
- *User-legitimators* (e.g., product champions, celebrities) shape the values and worldviews of other consumers.
- *User-consumers* adopt products and services, test new systems, and promote sociotechnical systems that support certain consumption practices over others.

Perhaps one of the most important emergent perspectives is the notion of "prosumers." Prosumers are actively involved in both the consumption and production of energy.[8] These people represent the new wave of decentralized energy use that relies on smart meters and alternative energy sources to generate electricity to support home energy management systems, energy storage, electric vehicles, and electric vehicle-to-grid (V2G) systems. Some of these prosumers—such as the farmers that invest in wind turbines in Denmark—are more producers than consumers. Smart prosumer-centered grids alter a number of fundamental attributes of conventional grids and the consumer connection. Smart grids enable sophisticated energy management capabilities that center on network metering or smart meters. Smart systems offer dynamic pricing possibilities and are built to accommodate distributed generation. Additionally, prosumer-centered smart grids can incorporate various types of storage (batteries, appliances, and cars) that enhance contributions from stochastic power flows like wind and solar power. Indeed, in an attempt to capture some of these benefits, many countries have begun to embrace far-reaching reforms of conventional systems, and policies are already under way to encourage the technological innovations necessary to cope with increasing amounts of power from intermittent sources and independent producers.[9]

Fundamentally, a prosumer market model frames the role of the customer in a more participatory manner (see table 7.1). In a prosumer model, consumers don't just respond to price signals by modifying consumption; they also modify supply-side behavior by altering investment patterns, and in doing so, they influence the market for services from electric utilities, transmission systems operators, or other prosumers. Prosuming can enable consumers of energy to save money while also contributing to wider social benefits by diversifying the energy supply and lowering greenhouse gas emissions from the electricity system and private transportation.

CONSUMPTION PRACTICES, INNOVATION, AND NORMS

In confronting any addiction, it is important to identify triggers—events or conditions that tend to catalyze or enable adverse behavior. The theory goes, if we can eliminate or at least reduce exposure to these triggers, then the addict is better positioned to resist the addiction. In keeping with our premise that how we use energy is a form of addiction, it serves a purpose to identify the influences that underpin energy usage patterns and habits.

TABLE 7.1. Comparing conventional electricity consumers and smart prosumers

Dimension	Conventional grid consumers	Smart prosumers
Resilience and response	Operators respond to clear signs of a problem. The focus is on a reaction and protection of assets following system faults.	Consumers or their devices can automatically detect and respond to actual and emerging transmission and distribution problems in real time. The focus is on proactive responsiveness.
Information and consumer involvement	Consumers are uninformed and nonparticipative in the power system.	Consumers are informed, involved, and active.
Nature of energy services	Services are produced in bulk, typically through a centralized supply.	Services are more modular and tailored to specific end uses, which can vary in quality.
Diversification	System relies on large centralized generating units where energy storage becomes a major capital cost. Energy providers tend to be large entities that often enjoy market monopolies.	System encourages large-scale distributed generation deployed to complement decentralized storage options, such as electric vehicles and stationary storage, with more focus on access and interconnection to renewables and vehicle-to-grid systems.
Competitive markets	Limited wholesale markets are still working to find the best operating models. They are not well suited to handling congestion or integrating with each other.	More efficient wholesale market operations are in place, with integrated reliability coordinators, high-tech load balancing systems, and minimal transmission congestion and constraints.
Optimization and efficiency	There is limited integration of partial operational data and adoption of regularly scheduled maintenance.	There is greatly expanded sensing and measurement of grid conditions, and technologies are deeply integrated with asset management processes and condition-based maintenance.

Source: Y. Parag and B. K. Sovacool, "Electricity Market Design for the Prosumer Era," *Nature Energy* 16032 (2016): 1–6.

This approach takes us into the realm of innovation theory, sociotechnical systems theory, and path dependence theory.

Innovators and Early Adopters

The pace at which innovations are adopted by consumers (or prosumers) will help to determine how quickly we can expect a transition to a new low-carbon, smart-grid energy system. Therefore, we need to understand the technological adoption process.

For almost a century, scholars have sought to understand the process by which innovations are diffused and ultimately adopted (or rejected) by their intended audiences. Influential here has been the work of Rogers and his book, *Diffusion of Innovations*.[10] At the core of this work is a framework for understanding the trajectory of innovations from adoption to mass market consumption. Rogers depicts the process as a normally distributed graph when plotted using adoption rates (see figure 7.1). According to Rogers, the adoption process entails five categories of users or adopters: innovators, early adopters, early majority, late majority, and laggards.

Rogers describes a four-part process through which an innovation is adopted: "(1) an innovation, (2) is communicated through certain channels, (3) over time, (4) among members of a social system."[11] Accordingly, understanding the catalysts in this process entails a four-part investigation.

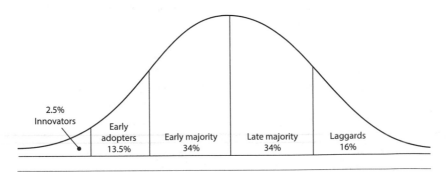

FIGURE 7.1 Diffusion of innovations curve (E. M. Rogers, *Diffusion of Innovations*, 5th ed. [New York: Simon and Schuster, 2003.

First, the perceived benefits associated with the innovation itself must be evident. The main premise is that an innovation must provide a superior portfolio of consumer benefits to usurp competitive offerings. If the portfolio of benefits is only marginally superior, the innovation might not be adopted by the mass market—it might become a niche solution adopted by only a segment of the market. Second, an array of communication channels facilitates the dissemination of information to consumers and public officials throughout the diffusion process. For product marketers, the reach of the communication channels through which the benefits of the innovation can be conveyed to consumers is a key factor behind expedient, mass market adoption. Today, thanks to the power of social media, the benefits of innovative products and practices can be conveyed in a viral manner; a recent example was the winter 2017 explosion of interest in crypto-currencies thanks to the spectacular market performance of Bitcoin.

Third, time can be a friend or foe to emerging technologies. If adoption patterns entail high switching costs, adoption can be hindered, giving competitors time to respond with new competitive offerings that can reduce market adoption. Delays in the launch of new mobile phone models can undermine a firm's bottom line for the entire year. Playing catch-up is tough business. Finally, we need to understand the socioeconomic context into which an innovation is being introduced in order to predict the pace and scale of adoption. For example, in many households in North America, automobiles are more than transportation tools. The choice of automobile is influenced by the worldviews of the owner. Picture what an owner of a Ford Mustang might look like. Now picture the owner of a Nissan Leaf. What does your own perspective on owners of these vehicles tell you about possible barriers to electric vehicle (EV) adoption?

Several studies have applied insights from literature on the diffusion of innovations specifically to energy use (particularly solar energy). In their review of the field, Wolske and colleagues[12] note that early adopters of solar thermal systems were more likely to see solar thermal technologies as less financially and less socially risky, less complex, and more compatible with their personal values when compared to members of the general population. Follow-up work in the United Kingdom suggests that, compared to early adopters, potential "early majority" adopters of residential solar photovoltaic (PV) systems needed convincing that residential PV systems are attractive, affordable, and hassle-free.[13] In the United States, research in California has found that peer influence matters. When solar adoption is more prominent in a particular zip code, the likelihood of a neighboring household adopting solar increases.[14] Similarly, a study in Texas suggests that the decision-making process for considering solar is shortened when homeowners either see solar panels in their neighborhood or talk to a neighbor who has solar.[15]

Individuals Constrained by Sociotechnical Systems

A second, very different perspective sees users not as independent adopters of innovations, but instead as unwitting pawns within existing sociotechnical regimes, or large technical systems, that constrain (rather than enable) innovation. According to this view, large sociotechnical systems such as electricity grids, transport systems, or telecommunications networks reflect "deep structures" in society that supersede natural geography or politics as key arbiters of societal change.[16] These systems weave together technologies and organizations, institutional governance systems, market structures, and cultural values. For instance, from a technological lens, a pipeline is just a physical conduit for oil or gas, but as part of a sociotechnical system it is an integral part of a broader network of pumping stations, operators, embedded knowledge, financial institutions, investors, landowners, import and export terminals, oil and natural gas refineries, energy traders, and consumers. Viewed from a purely technological perspective, a car is just a clunky box with an engine and wheels, but seen as part of a bigger sociotechnical system, it belongs to a network that includes roads, traffic signals, traffic laws, fuel stations and refineries, car repair shops, parts makers, licensing offices, insurance companies, and police and legal networks. This perspective suggests that change can be damaging to a broader number of stakeholders than just the users of a given technology. In order to understand impact, we need to shift our focus from individual technologies to the broader scope of "systems of systems."[17]

Sociotechnical systems crystallize and become resistant to change over time. If you think this perspective is overstated, you perhaps did not witness the 2008 Republican National Convention, where a room full of elected politicians and supporters took up an extended chant of "drill, baby, drill" in support of fossil fuel exploration. Thomas Hughes described such a manifestation as "momentum," or as a mass of "machines, devices, structures" and "business concerns, government agencies, professional societies, educational institutions and other organizations" that "have a perceptible rate of growth or velocity."[18] Bernward Joerges termed it "dynamic inertia," while others call it "lock-in."[19] *Momentum* and *dynamic inertia* describe a large sociotechnical system's propensity to continue along a given path due to the actions of numerous stakeholders, such as the work of educational and regulatory institutions, the investment of billions of dollars in equipment, and the work and culture of people working within an industry.

Stakeholders who benefit from incumbent technological regimes understandably prefer to discourage change. While users of incumbent technologies may seek increased efficiencies and profits, they do not want to see the introduction of new and disruptive technologies.[20] Habit can be a powerful supporter of the status quo and a barrier to change when behaviors are repetitive, as is the case with managing household thermostats, the commute to work, and diet.[21] Sadly for proponents of alternative energy technologies, this fact poses a dilemma because a great many affluent individuals and firms benefit from existing technologies resourced by fossil fuels. As Hughes points out, it takes disruptive events, such as war, to dislodge dynamic inertia. Fortunately, as we saw in chapter 1, the new economics of energy markets and the threat of climate change represent two significant developments that possess the power to disrupt.

Arbiters of Social Practice

A third more deterministic view sees consumers as agents that are influenced by social practices. Elizabeth Shove and colleagues state that attempts to explain "why people do what they do" benefit from an understanding that behaviors are driven by beliefs, values, lifestyles, and tastes that express personal choice.[22] Theories of social practice are recursive in the sense that they suggest that human action and social structure are mutually co-constructed.

Sociologists have argued that, under this lens, social practices can be better understood by dividing a given practice into four dimensions, described in table 7.2.[23] In the third column of the table, we relate this perspective to our theme of energy transitions.

This table illustrates that what people do is never simply reducible to utilitarian or purely rational choice. Instead, consuming something is a performance. If the technology does not align with the type of performance the consumer desires to put on, adoption will be thwarted. Conversely, materialities can come to mutually align, depend upon, or support each other, creating "systems" or "circuits" of practice. These circuits of practice can

TABLE 7.2. Four dimensions of human consumption and the connection to energy transitions

Dimension	Description	Example of relevance to energy transitions
Materialities	Include tangible physical entities, technologies, and the material objects that can convey meaning when seen.	An electric vehicle (EV) parked in your driveway is a symbol of choice, and rooftop PV panels are a type of "green bling."
Competencies	Include skills, habits, knowledge, tacit knowledge, and techniques.	Consumption choice tells the viewer what you know about the world (i.e., that you are aware of climate change and you are comfortable with new technologies).
Meanings	Include ideas, symbolism, aspirations, and other cognitive dimensions.	Consumption choice also gives the viewer insight into your worldviews (i.e., you wish to be a difference maker).
Connections	Describe how certain practices emerge, persist, shift, or disappear over time.	Continued diffusion of the EV depends on the link between recharging infrastructure and urban lifestyles.

Source: E. Shove, M. Pantzar, and M. Watson, *The Dynamics of Social Practice: Everyday Life and How It Changes* (London: Sage, 2012).

either empower or constrain technological diffusion. The last dimension in table 7.2 (connections) underscores the idea that practices can change over time.[24] All that is needed is a technology portfolio that better integrates with a consumer's competencies and the message that adoption of the technology is meant to convey.

Agents Expressing Values and Norms

A final perspective—values, beliefs, norms theory, or VBN—sees users not as innovators, unwitting agents, or arbiters of social practice, but as people merely trying to express individual values and norms. This approach comes from the domains of psychology and environmental sociology. Jackson called it a "theory of pro-environmental consumer behavior."[25]

VBN researchers attempt to describe why consumers make choices about the technologies or environmental practices they adopt. The theory serves as a strong critique of rational actor theory, arguing that decisions are often not based solely on reasoned analysis. As Dietz[26] and Dietz and Stern[27] have noted (and Macey and Brown[28] have shown empirically), many consequential actions are the result of habits and have complex motives that do not always reflect a reasoned assessment of trade-offs. Those who are addicted to constantly checking e-mail inboxes should immediately relate to this idea. It tells us that rational appeals alone might not be sufficient to stimulate changes in consumer behavior.

Moreover, the influence of values on decision making is often constrained by what is practical or satisficing. One might wish to reduce personal greenhouse gas emissions by reducing air travel, but the demands of one's vocation make it hard not to fly. A more sustainable household energy solution might be to erect solar panels on one's home; but one's financial status might instead favor the adoption of low-energy lighting or enhanced energy conservation.

At the core of VBN theory is the influence of "values" and "norms." Unlike a preference, a value is a general, unwavering principle that causes an actor to prioritize one thing over another. Values provide "a standard for assessing our behavior and that of others."[29] Core values rarely change over short time spans, and when they become a factor in a consumption decision, the importance of all other variables tends to become deemphasized. A good illustration of the influence of values on energy governance can be found in nuclear power politics. In many cases, the debate over whether to adopt nuclear power comes down to competing perspectives on risk and the ethics of exposing others to risk caused by the choices that those in power make.[30]

Values can be centered on environmental governance, or on environmental ethics, such as preserving an affinity with nature or coexisting with other species; they can be humanistic, such as aspiring for social justice or equity for all; and they may even be self-determining, such as a desire to protect one's family or to respect one's elders.[31] VBN theory posits that values affect beliefs, which in turn affect personal norms and action in a sequential fashion, as figure 7.2 summarizes.

This perspective is important because agents of change face a high degree of sociotechnological, economic, and political resistance to change. In order to break these shackles that inhibit change and foster an expedient transition, those in favor of a transition will need to find a way to enhance collaborations. Indeed, recent literature has shown how family, social networks, community systems, and institutions can influence personal norms, helping to explain the value-action gap that has been so dominant in the field of clean energy.[32] The role of key external "referents" harkens back to the work of Ajzen and Fishbein, who distinguished between social and

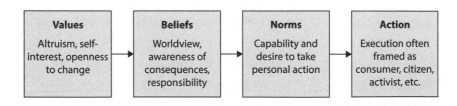

FIGURE 7.2 Values-beliefs-norms theory chain of causal influence (T. Dietz, "Environmental Value," in *Handbook of Value: Perspectives from Economics, Neuroscience, Philosophy, Psychology and Sociology*, ed. T. Brosch and D. Sander [Oxford: Oxford University Press, 2015], 329–349.)

personal norms in their theory of reasoned action.[33] The trouble is, with such diversity among individuals, comrades in arms are not easy to find.

VBN theory aligns with advocacy coalition theory, which suggests that people hold core values that inhibit change because the status quo represents an aggregation of human preferences and power.[34] Often the connection between core values and energy choices is not real, but perceived. For example, consider the phrase "freedom and transportation." What comes to mind? To most Americans this phrase conjures up an image of a convertible sailing along a vast highway. For people who consider freedom to be a core value, the symbolic connection to private automobile ownership makes the promotion of mass transit a threat to a core value—freedom. Advocacy coalition theory suggests that efforts to build coalitions for change need to reframe such artificial connections in order to dislodge the core value from the object to be changed. For example, if the core value of "freedom" is seen as influential in a campaign to reduce private automobile ownership, advocates of change might wish to reframe car sharing as a more accurate enabler of freedom—cars available whenever and wherever one wishes, choice of a different model of car each time, and freedom to use one's savings to purchase other goods or services that enable freedom (e.g., an annual trip).

Synthesis

All these perspectives on consumer behavior yield insights into creating strategies that could possibly expedite an energy transition. The innovation studies literature tells us that patterns of adoption are important, as are the signals that peers in a given community can give to others. Adoption rates can be amplified through better use of media and more effective strategies to convey the added value of emergent technologies. Coalitions that seek to foster the energy transition can leverage the impact of peer pressure to expedite trial and adoption.

The sociotechnical systems literature emphasizes the importance of viewing change from a systematic perspective. For an existing network to support change, the detrimental impact from change needs to be offset by benefits for the vast majority of stakeholders (particularly those in power), not just for the ultimate consumers.

The work of the social practice theorists details the importance of both conscious and subconscious routines, habits, and rituals that lock in technological regimes. To enable change, proponents need to understand that technologies are part of a broader social world composed of interest groups with wide-ranging views of how the technologies might fit in.

Conversely, VBN theory suggests that there are also internal drivers that influence technological adoption. In order to foster change, one needs to reframe change as a way to better encapsulate one's ideologies and values. Smart grids and low-carbon fuels must be positioned as enablers of self-realization. It is as important to convey what change signifies as it is to convey what economic or utilitarian benefits arise from change. Figure 7.3 summarizes many of these challenges—highlighting that we are indeed dealing with change amid a dynamic system with numerous moving parts.[35]

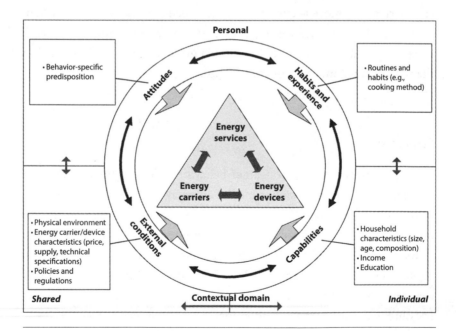

FIGURE 7.3 An integrated visualization of energy user behavior (Reza Kowsari and Hisham Zerriffi, "Three Dimensional Energy Profile: A Conceptual Framework for Assessing Household Energy Use," *Energy Policy* 39, no. 12 [2011]: 7514.)

SOME PERSISTENT BARRIERS TO CONSUMER CHANGE

Because of the complex influences on consumer behavior, it should come as no surprise that a complex mix of barriers impedes attempts to encourage consumers to become more energy conscious and adopt more sustainable behavior. Here, we focus on four interconnected sets of obstacles: apathy and ambivalence, ignorance and misunderstanding, culture and customs, and active resistance.

Apathy and Ambivalence

Over time, humans have transitioned away from kinetic energy (natural energy flows harnessed through simple technologies such as using sail cloth to capture wind) to chemical energy (usable energy stored in a form that can be stockpiled, transported, and released in concentrated form whenever and wherever humans desire it).[36]

Moreover, the role of the consumer in the production process has declined. In the 1800s, for example, most homes were heated by wood stoves, meaning that the person heating the home was often also the one chopping and stacking wood. This form of household heating was very labor-intensive, since wood had to be retrieved and loaded frequently. Then along came the coal-fired furnace, which reduced time spent on energy production because coal was more energy-intensive and it arrived at the consumer's door through more elaborate supply-chain models. Yet users still had to shovel coal into the furnace, and the daily ritual of loading coal made the householder aware of dwindling fuel supplies. The oil-fired furnace, which came next, eliminated the requirements of daily tending through the development of storage tanks. Finally, the modern-day natural gas or electric furnace has removed almost all human involvement in the production of heat apart from flicking a switch. Fuel collection, distribution, and delivery (along with energy use) have become effortless and invisible.[37]

Humans have become disconnected from the process of generating energy. Access to energy is taken for granted, and the systems that provide electricity have, by and large, operated reliably enough to sow seeds of apathy and ambivalence. Want to dampen spirits at a party? Try talking about the wonders of modern electricity. In the United States, the physical distancing of power stations from most cities and neighborhoods exacerbates public apathy. U.S. energy expert Mike Pasqualetti writes:

> An out of sight, out of mind pattern misleads the public by suggesting that the environmental costs of electricity are less than they actually are. . . . As distance, technology, and our urbanized lifestyle came to cushion us from the direct environmental costs of energy, we become increasingly less aware and eventually less tolerant of the intrusions of energy development on our personal space.[38]

In other words, even though electricity generation places extreme demands on natural resources, the environment, and the marketplace, the decision to sequester mega power plants away from the public eye has removed energy provision strategies from the public agenda in many regions.

There is an out of sight, out of mind mentality at play here. When energy infrastructure or technology is invasive and visible, people are more aware of its role in society. This is why people tend to overestimate the energy consumed by household lighting (which is visible through fixtures) but underestimate the greater amount used by water heaters (where energy is consumed deep within the protected bowels of an oft-concealed heating unit). A survey of four hundred households in Michigan found that the average resident believed, wrongly, that he or she could save twice as much money reducing lighting than using less hot water.[39] Building contractors report that it is easier to sell a new home with solar collectors mounted on the roof than a home with passive solar design, added insulation, and other less visible features, even though the latter actions would save energy more cost-effectively. Electric vehicles are the darlings of clean energy; yet few owners of such vehicles question where the electricity that powers their cars comes from. With coal-fired power plants being the dominant form of electricity generation and greenhouse gas emissions from such factories being the dominant source of greenhouse gases, many owners of EVs today just might find that, upon deeper analysis, their efficient EVs are not necessarily the best way to reduce greenhouse gas emissions. A better way would be to shutter the coal-fired power plants.

Most people do not even consider electricity as a human-made technology. Once technological landscapes are in place, people fold them so completely into their psyches that those very landscapes become almost invisible.[40] In today's culture, most people conceive of technology as constituting the latest high-tech items—new iPhones, fitbands, and so forth. Inventions of far larger historical significance, such as electricity, no longer "count" as technology. Paul Edwards quips that "the most salient characteristic of technology in the modern (industrial and post-industrial) world is the degree to which most technology is not salient for most people, most of the time."[41]

Thus, many consumers of energy are, at best, ambivalent concerning new, low-carbon technologies. In Europe, for example, Chris Groves and colleagues argue that consumer segments are not uniform; different people have different cultural interpretations of the smart grid, smart meters, the smart city, and other visions of "smartness."[42] In consumer interviews, they observed that many users express an ambivalent attitude toward the value, service, and learning opportunities that smart energy systems may provide. Such ambivalence is especially strong among elderly participants, who tend to share a cornucopian narrative that is influenced by post–World War II abundance and the emotional, material, and symbolic benefits of energy, still viewed in contrast to the hardships of war-time scarcity.

Liddell notes that this combination of resistance, ambivalence, and other factors likely explains why it is so difficult to translate smart-meter adoption into efficiency savings.[43] The most successful reductions in consumption require a change in lifestyle and sustained "vigilance" that many householders do not possess or want to possess.

Ignorance and Misunderstanding

Apathy and ambivalence have sired two offspring—ignorance and misunderstanding. To give an example of the lack of knowledge regarding electricity, a comprehensive study undertaken by Southern California Edison, surveying thousands of consumers, asked: "Where does electricity come from?" Most people said, "Out of the plug in the wall," while others even said "lightning" and "static electricity." One of the authors of the study concluded that people in the United States have no conception of how electricity is generated or transmitted.[44] Surveys conducted by the National Environmental Education and Training Foundation have consistently found that more than half of Americans believe that hydroelectricity, solar panels, and nuclear power are the "principal" sources of energy for the country. Of course, the correct answer is fossil fuels.[45] Almost half (41 percent) of respondents in a Kentucky survey identified coal, oil, and iron as "renewable resources."[46] Another survey of American electricity consumers found that four-fifths were unable to name a single source of renewable energy (even including dams and hydroelectricity).[47] A separate study found that nearly 70 percent of flexible fuel vehicle owners (people who purchased automobiles that could run on gasoline and/or ethanol) were unaware that they were driving one.[48]

Lack of knowledge also prevails outside the Land of Trump. In energy-enlightened Denmark, more than two-thirds of Danish respondents did not know how much electricity an average Danish house uses, and about 85 percent of business leaders did not know how much electricity the typical Danish company consumes.[49] More than 63 percent of household respondents and 85 percent of industrial respondents did not know how much electricity cost them per kWh, and about 30 percent of respondents did not know which devices used the most energy inside a typical home. Indeed, fewer than 11 percent of business respondents and fewer than 16 percent of household respondents answered at least four of five energy literacy questions correctly, and fewer than 4 percent answered all of the questions correctly. Clearly there is a notable difference between ignorance of electricity prices and the belief that electricity comes from a plug in the wall, but the commonality is that failure to understand the nature of energy and its cost to society makes it hard to justify transitional investment decisions.

Lack of knowledge abounds in global energy markets. In China, one-third of respondents to a household survey indicated that they did not remember or care about the amount on their electricity bill.[50] In Papua New Guinea, a school principal believed that a small solar home system could produce enough energy to power a computer lab, a copy machine, lights in every classroom, and a range of appliances—when in fact it could only light four lamps.[51] In Bangladesh, a family that one of us (Sovacool) interviewed thought they needed to take down their solar home system and take it "for a walk" every day so it "wouldn't get tired." Another family thought they had to "add soap" to their waste to "make it clean" before it would work in a biogas digester.[52] In Mongolia, one household thought their solar system would work indoors and had positioned it inside their ger (a portable home structure) as a coffee table.[53] In Nepal, one mother thought she needed to cover her solar home system with leaves to make it "part of nature." Another thought it could "dry laundry" and placed socks and underwear on it.[54]

Such gaps in awareness directly contribute to consumers making ineffi-cient choices. For example, researchers from Vanderbilt University found that if people simply stopped idling their vehicles—keeping them "on" to get them "warmed up" while stopped at traffic signals and during tempo-rary stops—the United States could save 10.6 *billion* gallons (40.1 billion liters) of fuel each year and displace 99 million tons of carbon dioxide, off-setting 1.6 percent of all national emissions.[55] Similarly, if drivers in the United States learned to properly inflate their tires, they would save more crude oil than the maximum amount that could be annually produced from the Arctic National Wildlife Refuge in Alaska—more than 540,000 barrels of oil per day.[56]

Culture and Customs

If apathy, ambivalence, ignorance, and misunderstanding were not daunt-ing enough, cultural barriers (customs, routines, and collective norms) also impede low-carbon transitions in spectacular fashion. In most indus-trialized nations, many consumers believe that they are entitled to more energy-intensive standards of living, and many utilities believe it is their duty to give it to them at the lowest cost possible. For most of the past century, many consumers in industrialized nations have come to expect unbridled material progress linked to visions of a high-technology soci-ety and all the trappings that this suggests—new cars, large modern homes, and the accumulation of energy-intensive appliances.[57] Underpinning the realization of such visions is the abundance of cheap, reliable energy. Consumers have become conditioned into thinking that they are "endowed with an abundance of domestic sources of energy and (unfettered) access to foreign sources," and they continue to expect supplies to be simultaneously "never-ending and cheap."[58] Abundance affects the way such consumers use energy, how businesses develop and market energy-consuming devices, and how government establishes policies to govern the sector. Consumers have culturally shifted from a system premised on self-restraint and moderation in the 1930s and 1940s to the idea that freedom and social well-being are best achieved through sheer abundance and limitless consumption.[59]

These values, far from floating "out there" in the minds of people, become a powerful and socializing cultural force manifesting itself in what we come to define as the physical environment. One extensive, two-year study involving a national survey of 1,500 households and 125 public utilities in the United States found that all households have significantly constrained options concerning revisions to energy-related features within homes. Rigid architectural design, choice of appliances, and the presence of built-in equipment restrict change.[60] For example, central heating and cooling systems are big investments that allow people to move freely from one room to another without thinking about energy. However, these systems make it difficult to save fuel by closing off unused rooms. Architects design apartment and office buildings with windows that cannot open for safety, but such a design makes it impossible to incorporate natural heating and cooling. Visitors to Singapore and Hong Kong frequently comment on the chilly rooms they must endure as overcooling prevails in hotels, offices, and shopping centers.

Ironically, interviews with families about their decision-making processes show that energy consumption largely depends on the *nondecisions* people make. Families do not usually make explicit decisions on energy consumption. Rather, they engage in activities of their choice to meet particular goals and consume energy in the process. Families almost never decide how many gallons of fuel oil to burn in the next month or how many kilowatt hours of electricity to use. Yet family purchase decisions about other commodities, from sugar to shoes and molasses to movie tickets, are consciously planned. Thus, heating and electricity bills are the consequences of a family's lifestyle, rather than the other way around.[61]

In the developing world, even though the energy infrastructure (electricity grids, district heating systems, and established transport networks) is not as advanced as in industrialized nations, culture and customs also influence how households use (or misuse) energy services. One in-depth analysis of ten countries in the Asia Pacific region noted that cultural, social, and even religious attitudes can result in profligate and often careless use of energy and electricity.[62] For instance, in Bangladesh, an aversion to pigs has prevented predominantly Muslim households from adopting biogas units that would run on pig waste. Others simply refuse to purchase cookstoves because using biogas from excrement is seen as impure. In Nepal, many believe that hydroelectric facilities should serve the community for free. Because of this cultural stigma, there is little money to be made in charging rural households for pico-hydroelectricity. In Papua New Guinea, solar home systems have been hit by unusually high rates of vandalism, sabotage, and theft. Clans there share resources under a *wantok* system (where people are connected by strong social bonds) rooted in tribal traditions. Solar panels, which benefit particular households or individuals instead of the community, assault this *wantok* system. In Malaysia, air conditioning is omnipresent, with many businesses leaving their doors open to attract customers with the cool wafts of air that strike passersby. In rural Sri Lanka, many villagers associate grid electricity with deforestation and see it as a threat to community wood supplies. Many also see electricity as too risky because it causes lightning strikes and house fires.

Active Resistance

A final persistent consumer-side barrier is the active resistance that some users exhibit toward energy-saving technologies and practices. Psychological factors such as resistance to change, the desire for better control, lack of trust, and concerns over breach of privacy deeply shape attitudes toward the consumption of energy.

In the United States, for instance, freedom and control appear to exert a significant influence over energy choices. People will resist energy technologies that impede their freedom or appear to impinge upon their capacity to control their environment.[63] An army experiment utilized a device to save fuel by limiting the acceleration rate of cars or trucks. Drivers resisted the change, and in about 10 percent of the cases dismantled and disconnected the device themselves.[64] Researchers at the Center for Energy and Environmental Studies at Princeton University adopted a similar strategy to encourage energy efficiency in the home. There, researchers began by installing automatic thermostats to remove user control over the setting of room temperatures. The study found that people opposed the new devices because they did not have enough control over temperature settings. In a follow-on experiment to encourage user adoption of these automated controls, the Princeton team redesigned thermostats so that residents could temporarily override the system. This simple modification made the system more appealing to users and resulted in reduced household electricity consumption by 19.4 percent in the summer and reduced natural gas consumption by 31.3 percent in the winter.[65] The desire for freedom over one's choices is such a powerful driver of behavior that one psychologist has gone so far as to claim that the need to control one's personal environment "is an intrinsic necessity of life itself."[66]

In Europe, the desire for freedom over one's choices is complicated by perspectives of social justice. This situation has ironically led to resistance over the adoption of smart-home and smart-meter technology. During one of the trials for a smart-meter rollout in the United Kingdom, people "resisted" by refusing to utilize the meters properly. This resistance delayed the compilation of data and prompted others to drop out of the trial.[67] Two reasons appear to explain this resistance: the results of their efforts were seen as incremental, inconsequential, and therefore a waste of time; and some participants felt that the monitors put an unfair burden on households to take responsibility for carbon reduction compared to major energy consumers such as industry or government. Other studies have documented a similar "rejection of innovations" and feelings of disinterest and disenchantment arising from consumers in Finland[68] as well as other regions in the United Kingdom.[69]

In a comparison study of smart-meter perceptions in Europe, researchers found various manifestations of resistance couched in terms of invasion of privacy.[70] In focus groups, households saw the smart meter merely as an intrusion of power companies into their private lives.[71] One study cautioned that smart-meter technology can even be interpreted as an "information panopticon" that gives government or corporate entities insidious access to private consumer data, violating norms of data protection and security.[72]

Other consumers have been known to resist smartmeters for reasons of health. In a comparative assessment, one researcher noted that although privacy and security concerns remained paramount, those opposed to such devices expressed concerns over the health threats of wireless smart meters and the nonionizing electromagnetic fields that were attributed to such devices. The premise underpinning such concerns is that these devices transmit signals through microwave radiation, which might cause cancer or other afflictions.[73]

ENLISTING PROSUMER SUPPORT FOR LOW-CARBON TRANSITIONS

All is not bleak, however. There are empirical cases in which some of these user barriers have been reduced and consumers have adopted more sustainable lifestyles. Moreover, there is promise in the future merely because of the positive economics associated with change. Consider prosumers, who both consume and produce electricity. With the adoption of solar photovoltaic panels, smart meters, vehicle-to-grid automobile systems, home batteries, and other "smart" devices, prosuming offers the potential for electricity users and vehicle owners to reevaluate their energy practices for their own financial benefit. New household solutions, improved energy storage, declining costs, and favorable regulations have led to a steady increase in the number of consumers worldwide who are turning their homes into mini power plants. In the United Kingdom, for instance, almost 1 million households have installed solar panels, amounting to a total installed capacity of 8,700 MW.[74] This equates to about eight medium-sized conventional power plants. Smarter energy systems incorporating a variety of integrated energy management components are also becoming more widespread. These technologies enable consumers to optimize the electricity that they generate by matching it to needs and, when applicable, store the rest, thereby saving money or energy at various steps along the way. The global home automation market in 2014 was valued at around US$5 billion and was forecast to reach US$21 billion by 2020.[75]

As testament to the positive directions in which some consumers are now traveling, consider four interesting cases of consumer-led change that cut across countries, sectors, and platforms: Vandebron in the Netherlands, the Carbon Cooperative in the United Kingdom, Green Mountain Power in Vermont in the United States, and ethanol initiatives in Brazil.

Vandebron in the Netherlands

Established in 2014, Vandebron, a Dutch-based startup, provides an online peer-to-peer energy marketplace platform for renewable energy.[76] Using Vandebron, local renewable electricity generators sell their energy directly to households and businesses, with only a small flat subscription fee on both sides. This peer-to-peer platform allows producers to receive better rates for energy, while consumers gain certainty that they are supporting local renewable generation. As of 2017, there were roughly 100 energy producers listed on the Vanderbron website, supplying power to meet the demands of more than 100,000 households.[77]

Consider how the Vanderbron program erodes many of the barriers to consumer uptake outlined earlier. The venture connects local producers with consumers, fortifying a sense of community. It also helps educate consumers through peer-to-peer interactions that consumers trust.

The Carbon Cooperative in the United Kingdom

The Carbon Cooperative ("Co-op") is a local energy organization established in 2008 by residents in Greater Manchester. In cooperation with the housing specialist Urbanism, Environment and Design (URBED), residents carry out changes to their own houses and community buildings to reduce household energy usage by collaboratively providing capacity building, training, and access to discounted materials, services, and low-cost finance. The Carbon Co-op is a not-for-profit company, 100 percent owned and run by participating householders (members), and all profits are kept within the cooperative.[78]

The Carbon Co-op dedicates itself to developing communities of people who wish to improve homes in Greater Manchester to 2050 standards in a quicker, easier, and cheaper way through sharing experience, knowledge, and bulk purchases. Value is captured by creating a community of knowledge and action. The Carbon Co-op's aim is to bring together people with useful skills such as retrofitting, electronic expertise, web development, or programming in order to serve as knowledge resources for others interested in adopting these systems for their own homes.

An example of one of their initiatives is Eco Home Lab, where the Carbon Co-op is partnering with a project team to deliver open-source tools for managing energy, either to improve energy efficiency in the home or to better understand energy systems as a whole. Once again, the manner in which such initiatives attenuate the barriers discussed earlier should be clear. Through collaboration, barriers to trust and obstacles arising through lack of knowledge are overcome. The collaboration also takes on properties of a social event, thereby mitigating reluctance to change by employing peer support.

Voluntary co-ops such as this are not a perfect solution. One potential limitation of the Carbon Co-op's model is that it relies on people volunteering time and effort to share expertise and knowledge. However, such collaborative frameworks do represent part of the solution by providing a personalized conduit for filling gaps in skills, knowledge, and time.

Green Mountain Power in Vermont

Serving Vermont, a state in the New England region of the United States, Green Mountain Power is a local utility company that is known for its innovative energy programs. Over the past several years, the utility has initiated programs and services aimed at reducing dependence on the central grid and increasing participation from its customers. The power company offers insights and opportunities for customers to save money on their electric bills while also contributing to clean air and climate mitigation. Green Mountain Power's programs show consumers how to exploit the energy-environment "double dividend" by deploying cutting-edge innovations. Overall, these programs are estimated to have reduced carbon emissions by 116,000 metric tons in 2017—a reduction of almost one-half ton for each of the power company's 265,000 residential and business customers.[79]

One of the lynchpin innovations in this program is the Tesla Powerwall home battery system. Green Mountain Power was the first utility to collaborate with the Tesla Company to provide this energy storage system to its customers. With an enrollment fee of US$15 per month for ten years, customers can purchase the backup battery system, which is twinned with renewable energy systems. Consumers can now power home appliances and charge electric vehicles during expense peak-hour intervals without drawing from the grid.[80] These investments have helped customers move off the central grid for two hours each day, saving about US$275 per household each year.[81]

The energy storage program is coupled with a bundle of other programs that help customers invest in advanced heat pumps, smart and energy-saving appliances, electric vehicles, and community solar. In addition, the utility company provides free smart devices to monitor and control water usage, heat pump usage, and usage by other home electricity appliances. Furthermore, it offers electric vehicle charging programs—US$29.99 per month for unlimited off-peak charging and a US$600 EV purchasing rebate for eligible income households.[82] In addition, Green Mountain Power is currently implementing the first community rooftop solar arrays for local low-income households.[83]

Ethanol in Brazil

A final example of user- and stakeholder-centric low-carbon innovation comes from Brazil, where a participatory approach to biofuel development has incentivized experimentation and innovation.[84] There, biofuel firms have formed active partnerships with sugarcane manufacturers and farmers, automobile companies, gasoline stations, and electricity providers to close the production loop.

Brazil created its national ethanol program "Proálcool" in November 1975, with the objective of increasing domestic ethanol production and substituting ethanol for petroleum in conventional vehicles. Brazil not only surpassed the initial goals of its program in its first three years, but by 2005 the nation was the largest producer of ethanol in the world, producing about 16 billion liters of ethanol per year before being surpassed by the United States in mid-2000.

Proálcool was driven by two policy tools: a compulsory mandate that all gasoline had to have at least a 10 percent blend of ethanol, and a voluntary mandate that automobile manufacturers design vehicles to run on 100 percent ethanol. It did not specify a particular fuel source or feedstock, and stakeholders experimented with sugarcane, beets, cassava, potatoes, corn, wood, bamboo, vegetable residues, coconut, castor seed, cotton seed, and "any other feasible crop." Sugarcane emerged as the preferential feedstock.

The program was premised on policy experimentation and flexibility. The government did not impose technology requirements on producers, allowing firms to select their own production processes to meet targets. When public mistrust and suspicion arose in 1981 over high ethanol prices and poor performance of ethanol vehicles, the government worked with auto manufacturers to improve ethanol-powered engines, set a price cap on ethanol to stabilize prices, and ran an educational program to inform consumers as to why these policies were in place. The government supported a great number of small sugarcane plantations, owned and operated by small independent growers. Government programs and subsidies were initiated to support these smaller enterprises, offering long-term contracts and providing price guarantees.

Lastly, the Brazilian program was highly inclusive and participatory. Although working conditions on sugar plantations and ethanol distilleries were indeed strenuous, workers in São Paulo received wages that were more than 80 percent higher than the agricultural sector average. To ensure that the sugarcane industry would not lose access to laborers, the government provided workers with additional bonuses tied to the value of sugar and ethanol produced each year. Environmental groups were invited to provide feedback, and discussions resulted in the government mandating that rainforest land could not be used for new sugarcane plantations, processing facilities, or ethanol plants. Brazilian automakers were integrated into the process early on. Moreover, as ethanol production increased, partnerships were cemented between ethanol producers and electricity suppliers because sugarcane ethanol produces bagasse, a usable fuel for power plants.

An update on this case study demonstrates how market forces can still be hard to beat. The cost of ethanol production in Brazil has been rising because of inflation, Brazil has discovered oil deposits offshore, and the price of imported refined gasoline has dropped. As a result, millions of Brazilian motorists have switched to gasoline, causing the "biggest crisis in the history of ethanol." Even the best of behavioral programs can struggle with the perennial problem of volatile energy prices.[85]

CONCLUSION AND THOUGHTS FOR REFLECTION

Users play complicated roles in energy and low-carbon transitions. They can be at times obstacles to change, while in other instances, they can be active agents of innovation. No single theory or academic perspective can explain consumer behavior fully or adequately capture the complexity of technology adoption.

With this in mind, we offer two conclusions. First, users matter, and they can do more than merely consume energy services. However, they are not homogeneous. Engaging with and integrating users into the change process come with immense challenges. You can't teach an old dog new tricks, particularly when the old dog likes his old tricks, is too busy chasing cats to care about learning new tricks, and doesn't particularly trust his trainer. Energy consumers are indeed old dogs in many respects. Many consumers may harbor apathy and ambivalence toward change, and some may even subscribe to cultural norms or fall prey to routines and beliefs that make change intractable. Still others may actively resist new energy systems—some for good reason and others just for the sake of resisting change.

Second, users and energy consumers have complicated, sometimes convoluted, and even contradictory motivations. This means that users:

- Remain largely encumbered by sociotechnical systems that inhibit change
- Have motivations irreducible to a single factor, making them difficult to isolate and hindering collaboration
- Make irrational or at least nonmonetary decisions that are difficult to model
- Frequently behave in ways that are inconsistent with rational actor theories or the bulk of research approaches informed by positivism
- Possess dynamic views and practices that can change over time
- Often remain misinformed (holding information that may be inaccurate or biased) or uninformed (the "empty bucket" model that presumes users need only accurate information or knowledge to make sound choices)

Therefore, to expedite transitions by enlisting consumers as change agents, proponents of change will benefit from establishing strategies that encourage collaboration and peer-to-peer learning. Just as the purchase of a new car was a social event at the turn of the twentieth century and the purchase of a TV was a neighborhood highlight in the 1950s, the coming energy transition needs to be positioned as a social event—one that even laggards cannot bear to miss out on.

8

Minimizing Governance Barriers and Creating Polycentric Networks

To this point, we have examined a number of significant obstacles to expediting an energy transition. We explored R&D strategies, addressing financial challenges, curtailing opposition from fossil fuel interests, policy portfolios for driving change, and altering the fickle nature of consumers. Yet, the transition that we are talking about is a good thing, isn't it? It should be clear from earlier chapters that renewable energy technologies can deliver benefits to any nation, including lower negative externalities per unit of energy services delivered, more stable and predictable fuel prices, consumer savings, and fewer greenhouse gas (GHG) emissions.[1] Renewable energy technologies also use less water than conventional power plants, have better energy payback ratios, improve energy security by diversifying the fuel mix, and provide greater local employment and revenue.[2] Renewable power stations do not melt down, rely on hazardous and combustible fuels, or depend on a fuel cycle of mining or milling that must beat, drill, or leach fuels out of the earth.[3] When a serious nuclear accident such as Fukushima occurs, it is a national calamity; but when a solar power plant breaks down, it is just another sunny day.[4]

The benefits of this transition to national well-being beg the question: shouldn't the guardians of civic welfare—our respective governments—be doing more to ensure we can realize these benefits sooner rather than later? Perhaps, more to the point, why do some governments seem to be getting the point faster than others?

Even within nations, some jurisdictions are more proactive while others are laggards. In terms of wind power development, German leaders get it, with state leadership in Mecklenburg-West Pomerania and Schleswig Holstein blazing new trails. Wind power provides over 35 percent of the electricity in Mecklenburg-West Pomerania and about 30 percent in Schleswig Holstein.[5] Denmark gets it too. For a brief period in July 2015, the nation covered all its electricity needs through wind power and still had 40 percent left over to export to Norway, Sweden, and Germany.[6] On February 22, 2017, wind turbines provided all of Denmark's power needs for the entire day.[7] Conversely, leaders in ultra-high-tech Singapore don't see the value of such a transition (at least not yet), nor does the leadership just up the road from Singapore in Malaysia.[8]

One of the thorny complications about effective climate and energy policy is that it requires effective governance. Governments must design strategies to overcome barriers and prevent the "diffusion of responsibility" by getting all major stakeholders in a given locale to play a role in supporting change. Yet the process is akin to herding cats. The costs and benefits of improving energy security and fighting climate change are distributed inequitably, ranging from household inequity to cross-border inequality. The costs and benefits are also spread over time.[9] Effective energy governance at this point in time, therefore, presents a collective action problem for the ages.

This chapter focuses on unraveling the various threads of this governance challenge. It begins by defining governance (and distinguishing it from government or policy) and by explaining why energy transition and climate change present unique governance problems. It then discusses six interconnected and highly intractable energy governance issues: policy trade-offs, transboundary externalities, the political economy of trade and intellectual property, subsidies, the nature of wicked problems, and the rapidity of temporal and technological change. These obstacles, when interwoven with the market uncertainties discussed in chapter 4, infuse national policymaking bodies with the collywobbles, often creating an environment of policy paralysis and thereby forcing subnational change makers to take action in a less unified manner. When subnational change makers also fall prey to these forces, energy transitions stagnate. Therefore, the penultimate section discusses how polycentric governance approaches may resolve paralysis at the national level, and it offers five examples of success in this regard: Danish electricity policy, off-grid energy systems in Bangladesh, Chinese cookstoves, transport planning in Singapore, and building codes in California (in the United States). The conclusion draws from these cases to offer insights about transitions and policy more generally.

GRAPPLING WITH GOVERNANCE

Governance is not just a fancy term for "government." Governments are only *one* of the means through which governance can be achieved. Yet even governments cannot be considered unified entities in most respects. In most nations, there is a degree of independence on the part of nongovernmental actors, and such independence can result in markedly different policy approaches. Although most academic research in public policy and international relations still focuses primarily on what governments do, in recent decades many scholars have begun to address governance that occurs outside of formal governmental structures. International relations scholars have a particular interest in nongovernmental governance, given the absence of any prospect of a formal world government to govern global-scale issues.[10]

Table 8.1 summarizes the nuances in terminology, suggesting that governance in its most basic sense refers to the processes, systems, and actors involved in addressing collective problems that individuals and markets cannot solve for themselves.[11] Of course, the central provider of governance is government, which possesses the authority to make rules and the coercive power to back them up. But the private sector, civil society, financial institutions, and a variety of other organizations can participate in and often drive governance. Examples include the World Bank setting conditions on the loans that it gives to developing countries and environmental organizations such as the Environmental Defense Fund working with companies

TABLE 8.1 Conceptualizing governance, global governance, and global energy governance

Term	Definition
Governance	Any and all of the myriad ways in which groups of people attempt to solve collective action problems, deal with market failures, and ensure the provision of public goods
Global governance	Efforts to deal with the wide range of border-crossing issues involving multiple states and other actors from multiple parts of the world
Global energy governance	Making and enforcing rules to avoid the collective action problems related to energy at a scale beyond the nation-state

Source: B. K. Sovacool and A. E. Florini, "Examining the Complications of Climate and Energy Governance," *Journal of Energy and Natural Resources Law* 30, no. 3 (2012): 238.

to enhance corporate social responsibility programs. Around the world, financial investment funds are beginning to withdraw investments from the fossil fuel sector, with evidence that financial institutions controlling $5 trillion in worldwide assets are actively divesting fossil fuel investments from their funds—they're cashing in their chips from the hydrocarbon casino.[12]

Energy governance encompasses rule making and enforcement and aims to overcome the collective action problems related to energy supply and use. It involves the *processes* of setting agendas, negotiating, implementing, monitoring, and enforcing rules and agreements related to energy, as well as the *actors* connected to energy, including those attached to governments, nongovernmental organizations, civil society groups, corporations, and the general society. Each governance network can differ by geographic scale or by breadth of stakeholders involved. Thus, *global energy governance* refers to the rules and actors concerned with energy issues that have cross-border ramifications.

At the core of the governance challenge is what is known as the "public goods" problem.[13] Markets, despite extraordinary success in matching human desires for consumption with producers of goods and services, regularly fail to provide goods and services at socially optimal levels because the incentives of producers and consumers are misaligned. Even when a group of people share a desire for certain goods and services (such as security of property rights, clean air, or national defense), the free market might not effectively satisfy these needs. No one has an individual incentive to produce such goods because once they are produced, everyone benefits from them, even if they do not pay for them. Thus, the "consumers" can free-ride. "Producers," knowing the consumers will not pay, do not produce.

Responding to the public goods problem is a political process. First, some system is required to determine which public goods or services societies think are desirable to provide. This fact suggests that discourse between stakeholders needs to take place. Although one might be tempted to think that this process is characterized by town hall meetings, the process of mustering social consensus is much broader. For example, an evening news report on pollution levels in a certain city can be considered part of this discourse, as can chats between neighbors on the ills facing a neighborhood. Second, there has to be actionable agreement over who gets to make the rules and against whom the rules will be enforced—in other words, who constitutes the group to be included in the "collective" part of *collective action*. Third, agreement is needed regarding what rules ought to be put in place to have the public goods or services provided and who will pay what share of the cost of producing those goods. Fourth, a system needs to be implemented to manage and monitor the process of delivering the public good. At each step of the way, stakeholders compete to ensure that their vested interests are served. This suggests that the challenge of overcoming the public goods provision is highly politicized, with those with power and influence striving for a central position to control any challenges to their vested interests.

Most citizens are accustomed to assuming that the national government is best positioned to resolve public goods problems. After all, central governments are ideally situated to lay down the law and ensure an equitable distribution of burden (and benefits) in the provision of public goods. However, entrusting politicians to arbitrate over a political process is fraught with flaws. For one thing, political leaders are susceptible to coercion from vested interests. Moreover, resolving public commons challenges often requires short-term investments that lead to long-term gains. This is not the formula for success for politicians who wish to be reelected every two to four years.

Many national governments have little capacity to make and enforce rules to reflect the will of the people, and failures can have serious spillover effects on citizens of other lands. As the literature on "state failure" makes clear, the scores of countries whose governments are unable to provide basic public goods engender social dissent, as well as crime and pollution that sow disorder beyond their borders in such forms as refugee flows, terrorism, organized crime, and failure to collaborate effectively in international cooperative efforts.[14]

Globalization has both increased the demands on national governments and made it more difficult for them to govern. Citizens hold ever-higher expectations of what governments should provide (stability, prosperity, opportunity, security); at the same time, globalization reduces the capacity of national governments to control events within their own borders. In a globalized world, many issues require decision making across national boundaries, yet the short-term political structures found in most nations make cross-border governance extraordinarily difficult. In other words, many national public goods problems have become transnational public goods problems, and that portends complexity and amplified contentiousness.

SIX PESKY GOVERNANCE PROBLEMS

In the quest to provide reliable, affordable, and efficient access to energy services while making the transition to a low-carbon energy system without producing extraordinary disruptions and incurring untenable economic costs, stakeholders involved in the governance process will need to overcome at least six governance challenges.

Policy Trade-offs

One governance challenge concerns the competing priorities surrounding the achievement of energy security objectives. Encouraging demand-side improvements and enhanced energy efficiency can reduce peak generation levels on electric power grids but could also undermine the profitability of utilities. Therefore, one can understand the reluctance of privately owned utilities to promote energy conservation. Energy taxes can promote greater efficiency and reduce the amount of energy used, but such policies also impair energy sales. Again, one can understand why fossil fuel–producing nations are not fans of such policies. Establishing new standards for fuel efficiency can lower dependence on oil, but more rigorous standards impose costs on automobile manufacturers. If you were the CEO of Ford, what would your position be on this question? Creating a more efficient network of roads can lower congestion and improve automobile fuel economy, but it would also make it easier for people to get around, resulting in more kilometers traveled. Shifting from one reliable electricity source (such as hydroelectricity) to wind energy (intermittent and distributed) and natural gas (prone to volatile prices and GHG emission) would increase diversification, but it could worsen overall system dependability. This list of trade-offs is emblematic of how improving some aspects of energy security can undermine other aspects.

One study investigated five distinct strategic paths that are designed to prioritize different facets of energy security: (1) promote self-sufficiency, (2) provide energy services at the cheapest price possible, (3) decentralize expansion of electricity grids, (4) mitigate GHG emissions, and (5) foster energy systems that can operate under conditions of water stress and scarcity.[15] In comparing these five programs, as figure 8.1 indicates, there are more strategic differences than similarities. Some of the five outcomes benefit from similar strategies: both the GHG reduction and water governance paths emphasize a phaseout of coal; both the enhanced affordability and enhanced access paths promote shale gas; and both the oil curtailment and GHG reduction paths promote big hydro. However, there are even more examples of where the programs give rise to contradictory strategies. The GHG reduction program benefits from the expansion of nuclear power, whereas the water governance program benefits from less nuclear power capacity. A program that prioritizes affordability spurns taxes and subsidies, whereas a program that prioritizes enhanced access depends on these policy tools. A strategic path that stresses affordability frowns on taxes and

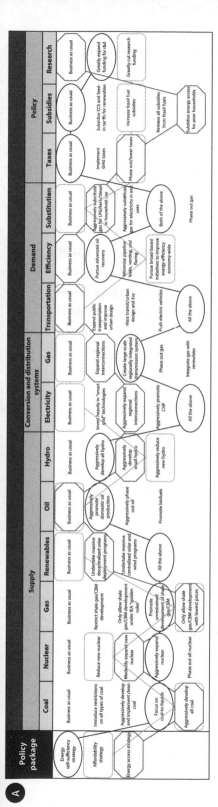
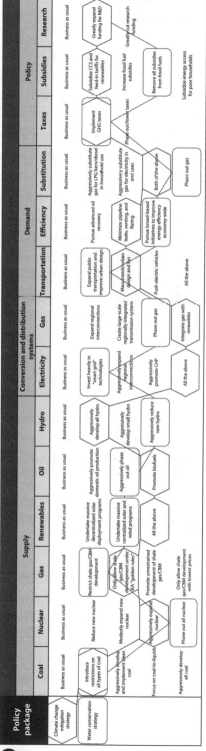

FIGURE 8.1 Competing policy packages for achieving energy security goals; a. conflicts between self-sufficiency, affordability, and energy access; b. conflicts between climate change mitigation and water conservation; CBM = coal-bed methane, CHP = combined heat and power, LPG = liquefied petroleum gas, CCS = carbon capture and sequestration (B. K. Sovacool and H. Saunders, "Competing Policy Packages and the Complexity of Energy Security," *Energy* 67 [2014]: 641–651.)

regulations, whereas both the GHG emissions reduction and water governance paths view taxes as a way to reduce environmental impact. No single technology package optimizes all of the energy security criteria. There will be winners and losers.

When multiple criteria are being used to define the notion of energy security, and as long as more than one person possesses a perspective on the topic, energy security can never be truly optimized. Energy security is in the eye of the beholder. It is optimized only when all parties can agree on which criteria should be prioritized and what the weighting should be for each dimension of energy. This, of course, is never the case in any community. Consequently, energy security planning is about managing trade-offs and risks and attempting to find common ground among a plethora of diverse stakeholders. The risks will never be fully eliminated, and the stakeholders will never be fully satisfied. In short, it will never be possible to provide policymakers with a definitive "laundry list" of policy prescriptions for achieving energy security.

This discussion suggests that managing policy trade-offs will always be a necessarily contentious process. The manner in which this task is undertaken can significantly influence both the acceptability of any given policy and the degree to which stakeholders are willing to seek compromise. Inclusivity reigns supreme when approaching this challenge.

Transboundary Externalities

Energy systems are at the heart of numerous transboundary problems. At the top of the list is climate change, but other examples include calamities such as nuclear meltdowns, oil spills, coal mine collapses, natural gas wellhead explosions, and dam breaches. There are also toxic pollutants such as mercury and phenomena such as acid rain that are side effects of the coal-fired power generation process. Such ills do not respect national borders and cause chronic disease, morbidity, mortalities, crop destruction, and damage to ecosystems. In nuclear power, there is the ongoing maintenance of caches of spent nuclear fuel, a common heritage issue because of the long-lived radioactive nature of high-level nuclear waste. If you are curious about why this might be a transboundary issue, it might be worthwhile to pay a visit to Kazakhstan to see the waste storage facilities for Russian nuclear waste, or to the Tokai waste-processing plant in Japan, which is the central clearing house for nuclear waste generated by plants around the country.

The transboundary impact of carbon dioxide emissions is extensive. One study analyzed trade data from 113 countries across 57 industry sectors and concluded that almost 25 percent of global emissions of carbon dioxide each year are from internationally traded commodities and products.[16] In some countries, such as France and the United Kingdom, the number exceeded 30 percent—that is, almost one-third of emissions were affiliated with imported products. Conversely, other nations such as China, India, Russia, and countries from the Middle East were net carbon exporters, as shown in figure 8.2. Follow-up research found that 37 percent of global emissions

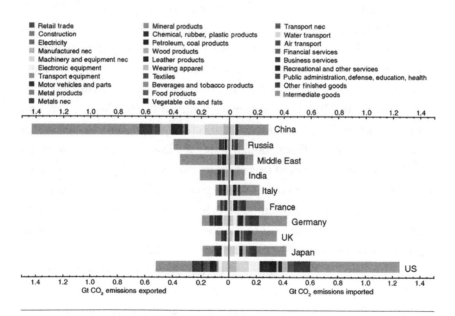

FIGURE 8.2 Balance of CO_2 emissions embodied in imports and exports of the largest net importing/exporting countries (and Middle East region); colors represent trade in finished goods by industry sector; traded intermediate goods (gray) are those used by industries in the importing country to meet consumer demand for domestic goods; "nec" refers to "not elsewhere classified" (Steven J. Davis and Ken Caldeira, "Consumption-Based Accounting of CO_2 Emissions," *PNAS* 107, no. 12 [2010]: 1–6.)

were from imported fossil fuel stocks—that is, they were not consumed in their countries of origin—and once again confirmed that 23 percent of global emissions were embodied in traded goods and spread across a supply chain involving at least two or more countries.[17]

Examples of transboundary pollution associated with energy generation also convey the staggering amount of damage that is exported across national borders. Approximately two-thirds of the anthropogenic mercury emissions generated in the United States is transported outside its borders, whereas 40 percent of the mercury falling on the country comes from elsewhere, mostly generated in Russia and the Asia Pacific region.[18] Possibly in retaliation for Canada's export of Justin Bieber to the United States, between 50 and 70 percent of Canada's acid rain comes from U.S. sources. Going the other way, 2 to 10 percent of America's pollution in this category comes from Canada.[19] Modelers from the National Oceanic and Atmospheric Administration's Pacific Marine Environmental Laboratory and GEOMAR Research Center for Marine Geosciences calculated how, because of ocean currents and the way that cesium 137 embeds itself in marine ecosystems, the radioactive fallout from the Fukushima accident in Japan was as much as ten times more dangerous when it reached the Pacific coast of North America in 2015.[20] Turkey, Syria, and Iran have all heavily dammed the headwaters that flow into the Tigris and Euphrates Rivers for electricity and irrigation, leaving the Shatt al Arab River without enough water for livestock, crops, and drinking. This situation has forced tens of thousands of Iraqi farmers to abandon their fields.[21] Forest fires in Indonesia, largely caused by slash-and-burn land clearing undertaken to accommodate palm oil plantations that provide the feedstock for biodiesel, give rise to transboundary hazes that increase hospital admissions and diminish rates of tourism in Malaysia and Singapore on a recurring basis.[22] An agreement to curtail these emissions—the ASEAN Agreement on Transboundary Haze Pollution—was finally ratified by Indonesia in 2014, twelve years after it was made available for ratification.

The Political Economy of Energy Trade and Intellectual Property

Meeting the growing global demand for energy will require massive investment—tens of trillions of dollars—with significant challenges stemming from anachronistic regulations, trade constraints, and the need to protect intellectual property rights.

Decisions about cross-border energy investment are influenced by numerous stakeholders, both private and public. Investments in cross-border energy infrastructure (and indeed many types of domestic infrastructure as well) are regulated primarily through a large, interlocking web of bilateral investment treaties (BITs), whose terms can actively discourage governments from making the regulatory changes needed to encourage the development of cleaner and less polluting systems. Many such treaties aim to protect the foreign investor from any financial loss arising as a result of adversarial state action. Those state actions include not only direct expropriation of property, but also the enactment of new laws or regulatory policies that force the investor to make changes that result in a loss. Thus, regulatory changes aimed at encouraging the development of clean energy could lead to expensive litigation. Under most BITs, such claims are settled by international arbitration panels that have generally favored investor interests.

The grand-daddy of all transboundary investment treaties is, of course, the World Trade Organization (WTO) framework, which sets down policies for governing international trade between the 164 member states. Although it was certainly not designed to undermine environmental governance, its policies have at times inadvertently done so. Global trade in energy fuels and services alone amounts to more than US$1 trillion a year. Since many energy exporters, including OPEC members, Central Asian countries, and Russia, are all members of the WTO, the forum serves as a useful arena for challenging policies that fossil fuel exporters deem undesirable. One of the key precepts of the WTO is that of most-favored-nation status. This principle mandates that member states must not enact policies that discriminate between trading partners.[23] Accordingly, when nations establish taxes and other disincentives to discourage fossil fuel use and the fossil fuel supplies are imported, a nation is potentially open to charges within the WTO of breaching this principle. Similarly, providing direct support to renewable energy industries may fall afoul of WTO prohibitions on subsidizing specific industries within a sector. Clearly there is a double standard at play here. Historically, in most nations, fossil fuel and nuclear firms have benefited from trillions of dollars in direct and indirect government subsidies, often defended by governments as necessary investments to ensure national competitiveness. Yet, when the same policy mindset is applied to renewable energy technologies, proponents of the status quo cry foul and engender alarmist critiques in the media that emphasize the economic perils of upsetting market dynamics and favoring less competitive technologies. This fact suggests one thing: investment treaties and market protectionism are susceptible to political influence. In other words, the industries that adopt the best political strategies stand to gain.

The defense of global intellectual property rights creates additional impediments to the diffusion of new energy technologies across countries. A debate has been raging in many industries over the importance of preserving intellectual property rights in order to spur innovation. Some argue that strong intellectual property right protections promote innovation, while others posit that strong patents deter innovation diffusion and raise prices. In the case of energy technologies, charges of intellectual property right infringement have been used to block entry into the wind turbine, solar panel, and hybrid electric vehicle markets in Japan and the United States and to prevent the acquisition of clean coal technology by Chinese firms.[24] Many countries in the developing world, moreover, do not own the intellectual property rights for the newest or most efficient energy systems, meaning that they have to license Western technology to break dependence on fossil fuels, particularly in markets such as China and India. One assessment of the barriers facing low-carbon energy technologies (such as clean coal and carbon capture and sequestration, nuclear power, energy efficiency, and renewable power plants) found that many firms were reluctant to sell new energy technologies in developing markets for fear that their intellectual property rights would not be respected and enforced by the WTO and other relevant authorities.[25]

Subsidies and a Cycle of Regulatory Addiction

National subsidies historically favor fossil fuels and nuclear energy, exacerbating the security and environmental governance challenges already described. The International Monetary Fund (IMF) estimated in 2015 that global fossil fuel subsidies amounted to US$5.3 trillion on a post-tax basis, equivalent to 6.5 percent of global GDP.[26] As figure 8.3 indicates, coal and petroleum received the lion's share of these subsidies. Nationally, the largest subsidies in absolute terms were in China (US$2.3 trillion), the United States (US$699 billion), and Russia (US$335 billion).[27] Subsidies such as these distort market prices and artificially create demand for both energy and its associated infrastructure. Developmentally, energy subsidies have heavily favored technologies that are the least efficient (from a thermodynamic standpoint) and most destructive to the environment, with the bulk of research subsidies going toward nuclear and fossil fuel systems.

In many industrialized countries, especially the United States, coal is the ultimate welfare case. Coal producers still receive a depletion allowance for mining operations, tax deductions for mining exploration and development costs, special capital gains treatment for coal and iron ore reserves, special tax deductions for mine closure and reclamation, research subsidies, and black lung compensatory benefits paid for by national governments. Oil too enjoys a bounty of freebees. Oil and gas producers receive similar depletion allowances, bonuses for enhanced oil recovery, tax reductions for drilling and development costs, fuel production credits, and research subsidies.

Although nuclear power is a very small player in the global energy mix, subsidies in support of this technology are also notable. Nuclear energy operators and manufacturers benefit from massive loan guarantees, research funds, public insurance and compensation against construction delays, tax breaks for decommissioning, tax credits for operation, and government-funded on-site security and off-site nuclear waste storage.[28] As we are also observing in Japan, nuclear power providers in Japan are also not held fully responsible for plant negligence; instead the government is picking up the tab. Subsidies such as these are not limited to rich countries. The world's poorer countries (members that are not part of the Organization for Economic Co-operation and Development [OECD]) subsidize oil exploration and production at more than US$90 billion a year.[29]

Subsidies have global energy governance ramifications in two senses. First, they mirror the impact of trade barriers and protectionist tariffs. Brazilian exporters, for example, face import tariffs that add at least

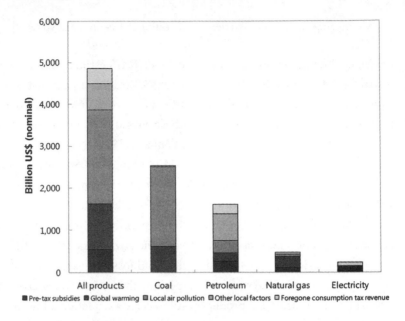

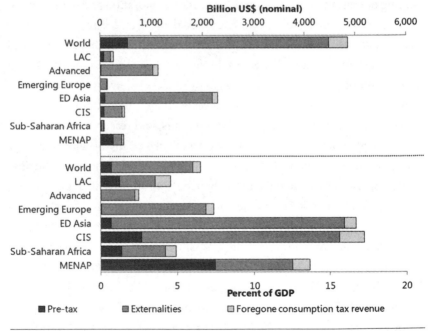

FIGURE 8.3 Global energy subsidies by sector/product and region; *top panel*: by sector/product; *bottom panel*: by region; CIS = Commonwealth of Independent States, ED Asia = emerging and developing Asia, LAC = Latin America and the Caribbean, MENAP = Middle East, North Africa, Afghanistan, and Pakistan (David Coady et al., *How Large Are Global Energy Subsidies?* IMF Working Paper WP/15/105, May 2015, p. 21, http://www.imf.org/external/pubs/ft/wp/2015/wp15105.pdf.)

25 percent to the cost of ethanol exported to the United States and that add more than 50 percent to the cost of exports to the European Union (EU).[30] Meanwhile, many of the same governments that apply tariffs to foreign imports exempt local production of biofuels from fuel excise and sales taxes. The artificial advantages these dual policies give to domestic suppliers is a possible violation of free-trade rules because foreign imports are not given the same exemption. However, challenging the process is like picking up dried peas with chopsticks—not always easy to pin down.

The second ramification of energy subsidies is that they tend to promote a subsidy arms race. Once one country subsidizes a particular technology or energy initiative, others are at a competitive disadvantage unless they follow suit.[31] Moreover, subsidies, once enacted, fuel powerful networks of vested interests, which make it difficult to claw back subsidies, and these higher prices cascade into the cost of other related materials. For example, subsidies for biofuels have resulted in higher prices for crops such as wheat, rice, soy, maize, and oilseeds, which have in turn caused the prices of staple foods and consumer products to rise. One study estimated that domestic energy subsidies for palm oil contributed to global price increases of maize by 60 percent.[32]

Importantly, subsidies become self-replicating because, once enacted, they engender the creation of long-lived infrastructure and capital stock. This situation justifies further subsidies to prop up the jobs created and to avert market disruptions. Coal and nuclear plants built forty years ago, for example, still receive subsidies for coal mining and uranium enrichment.[33] One study referred to this phenomenon as the energy subsidy "trap": once a government begins subsidizing, such efforts become protected and defended by beneficiaries.[34]

Subsidies also engender chain reactions. Once one country starts subsidizing a particular energy fuel or system, others are motivated to respond with their own subsidies to compete. Yet not all nations dole out the cash in similar amounts—some rule the roost when it comes to subsidies. Among the thirty industrialized nations forming the OECD—including the EU, Japan, Australia, and Korea—the United States is responsible for 70 percent of all subsidies for coal worldwide (when externalities are not included).[35] U.S. subsidies create higher demand for coal domestically and thus influence global prices, forcing other countries to subsidize their own energy sectors or force firms to pay higher energy costs.[36] Subsidies in this way create something very close to addiction.[37] Governments initially favor dispensation of market protections in exchange for political support, but they grow more dependent on the eggs from the golden goose that these subsidies create.

This trend of subsidy addiction is nicely illustrated by responses in developing countries to a crude oil price shock. In 2004, crude oil prices climbed to historic highs, increasing in price sevenfold before returning five years later to near 2004 levels. In the interim, the World Bank assessed the policy responses in forty-nine developing countries to these oil price increases.[38] The study found that many governments resorted to subsidies in the face of rising prices in an attempt to keep consumers protected. These new subsidies included credits for the exploration and production of oil, tax reductions, agricultural subsidies, discounts for passenger transport, price controls for fuel for fisheries, and partial compensation for domestic refineries. China, India, and Mexico alone adopted US$67 billion worth of new subsidies. However, the study noted that when prices for oil receded, the subsidies remained. Powerful constituencies had become dependent on them. Enacting subsidies is far easier than scaling them back.

Wicked Problems and Intergovernmental Regime Failure

Global energy and climate governance, like governance of other emerging transnational challenges, requires—get ready for the shocker—collaboration. When all the national actors are rowing in the same direction, international strategies can have substantial impact. A case in point is international progress in mitigating ozone depletion in our atmosphere. However, more often, attempts at international collaboration run into competing national interests, conflict over how the pie should be split up, or petty squabbles over the procedures and language used to frame agreements. Indeed, the major difference between life as a kindergarten student and life as an international policy negotiator is that kindergarten students enjoy the benefit of a teacher to intercede and resolve spats that have gotten out of control.

This situation poses several problems for those who seek national-level change in an area such as energy policy. First, as previously suggested, decisions made by intergovernmental organizations such as the WTO can constrain nations in developing policies that might serve domestic interests. For example, decisions to level the competitive playing field by incentivizing domestic firms to invest in clean energy can run afoul of international trade laws. However, it is the second problem that merits deeper examination. Decisions made within intergovernmental organizations are often so inefficient that critics can be forgiven for bandying phrases about such as *international regime failure*. It is useful to explore why this is the case.

Often intergovernmental efforts arise in response to an emergent problem of global concern—the collapse of fisheries, the ozone hole, and climate change being but three examples. Unfortunately, the problems are often tackled by specialized groups that are rarely fully inclusive and often are too narrowly focused to offer an effective systematic solution. For example, the global response to the oil cartel crises of the 1970s was to create an intergovernmental body—the International Energy Agency (IEA)—to coordinate the storage of three months' worth of oil. The theory was that this strategy would allow member states to draw on these reserves should the oil cartels once again restrict oil supplies, thereby buying the member states time to find a solution. The decision to focus on reducing the volatility of this finite resource artificially depressed oil prices for decades. If the IEA framework did not exist, we might already find ourselves in a world dominated by alternative energy. A second flaw of the IEA framework was that it excluded important energy consumers such as China and India. Therefore, when Chinese demand amplified at the turn of the century, the demand-side price shocks that ensued eclipsed the capacity of the IEA reserve standards to control prices.

When intergovernmental organizations could not adequately govern oil trade, private firms filled the void. This situation led to contractual structures (e.g., forward contracts) and oil market governance mechanisms that are more effective but also entirely private and, therefore, aimed at preserving vested interests. This problem has given rise to regulatory capture in many nations and made it far more difficult for nations to focus on what really matters—the creation of energy systems that are affordable, resilient, and accessible.

Even when intergovernmental efforts succeed, they usually do so in suboptimal ways. Seldom do international negotiations yield a consensus on direction and pace of change that exceeds the expectations of the majority of national players involved. National negotiators do not enter into negotiations with the mindset that they will seek to commit their nations to doing more than what they consider their fair share to be. Rather, the starting point is typically somewhere closer to doing the bare minimum. After heated negotiations, negotiators then strike agreement in the range of bare minimum plus a little bit.

When it comes to climate change negotiations and the question of how fast a transition out of carbon-intensive energy technologies should be, this *bare minimum plus a little bit* strategy is further undermined by the wicked nature of the problem. Wicked problems are problems that are characterized by uncertainty over how the problem came about, what the solution is, and how the problem will evolve due to the numerous causal variables involved and the complexity with which these variables interact. Uncertainty of this type prevents the identification of an objectively optimal outcome.[39] In addition, wicked problems tend to cut across various policy subsystems and governance levels, drawing in a wider swath of nongovernmental stakeholders. These groups seek to reframe problems and solutions, influence others through discursive deliberation, embed themselves in the global policy process, implement nongovernmental policy responses, and sometimes enact policy in parallel to or in opposition with government policy. In short, the wicked nature of the global energy challenge means that international negotiators have a hard time establishing what the bare minimum should be when approaching negotiations. As a result, opening negotiation positions tend to reflect a very flaccid contribution to transnational change.

For these reasons, many intergovernmental organizations that work on energy and climate change issues have been heavily criticized. The World Wide Fund for Nature (WWF) has criticized organizations such as the Forest Stewardship Council, the Marine Stewardship Council, and the Roundtable on Sustainable Palm Oil for having "limited" impact on conservation and development.[40] Various scholars have attacked the Forest

Stewardship Council for marginalizing local actors, failing to address the underlying causes behind increased consumption of forest products and deforestation, and exacerbating poverty.[41] Similarly, the Roundtable on Sustainable Palm Oil has been criticized for pushing industry interests over local stakeholder interests and for being plagued by noncompliance in regard to the degradation of peat forest governance.[42] The Roundtable on Sustainable Biofuels has been accused of facilitating land grabs— converting land needed by rural pastoralists or subsistence farmers into assets for elites.[43] Efforts to reduce emissions from deforestation and forest degradation in developing countries (REDD+) have been attacked for forcibly relocating forest dwellers without adequate due process and for worsening social, political, and gender inequality.[44] Finally, the Extractive Industries Transparency Initiative (EITI) has been similarly criticized for not facilitating meaningful improvements in governance in its first two candidate countries (Azerbaijan and Liberia),[45] for failing to facilitate reductions in poverty, for ongoing corruption, and for failing to deliver meaningful increases in economic development and investment.[46]

Temporality, Uncertainty, and Rapid Change

We live in dynamically evolving and increasingly unpredictable times. The first commercial text message was sent in 1992. A 2018 analysis estimates that there are now 65 billion text messages sent daily.[47] It took the radio thirty-eight years to reach 50 million people, television thirteen years, the Internet four years,[48] and Facebook only two years.[49] It has been estimated that there are 5.5 billion searches on Google alone each day,[50] and by 2018 there were more than 23 billion devices connected to the Internet[51]—that is three times more devices than people. With 326 million registered Twitter users, if Twitter were a country, it would be the third largest by population in the world after China and India, just edging out the United States.[52]

With new technologies come novel goods and services and different ways of doing business. Schools and universities today are preparing students for jobs that do not yet exist, that will employ technologies that have not yet been invented, to solve problems that we have not yet identified.

As an example of some of these problems we are only beginning to understand, global temperatures, extreme weather, and sea levels are all exhibiting accelerating rates of change. Consequently, the number of extreme weather and climate-related disasters is also increasing,[53] and the economic costs from these events are rising rapidly as a result of wealth and population increases.[54]

Rapid changes in technological innovation and resource availability cascade into a highly volatile economic picture. Price uncertainty, in turn, wreaks havoc on the governance of energy by confounding the analysis of energy security evolution. Consider how the inflationary pressures and price volatility affecting conventional energy prices, for instance, can alter analyses. When fuel stock prices vary in such an unpredictable manner, utilities and energy-intensive firms are forced to either take on more operational risk or try to hedge the risk through the purchase of forward contracts, which adds to the cost of doing business.

To summarize this section, complexity and rapid change are undermining the ability of humans to rationally manage energy systems, or at least rationally debate the competing forces that influence energy security. Complexity and change are corrosive to effective governance and cooperative efforts because change gives rise to arguments over whether an emergent trend is sustainable, subject to regression, or just a temporary detour along an otherwise linear path.

TOWARD POLYCENTRIC ENERGY AND CLIMATE NETWORKS

All is not lost, however. Some strategies and governance arrangements can overcome the obdurate challenges just discussed. One useful strategy for tackling complex, wicked problems is polycentrism. *Polycentrism*—or polycentric approaches to governance—refers to conditions in which citizens and interest groups organize into multiple layers of governance at various scales simultaneously. In one respect, polycentrism reflects a risk diversification strategy. If governance efforts fail at one level, efforts at other levels might be able to attenuate the failure. An example would be the ongoing inability of U.S. federal politicians to agree on support for wind power and how governance regimes in states such as Texas, Iowa, California, and Oklahoma have taken the baton and orchestrated growth in wind power capacity.

In another respect, polycentrism is an organic manifestation of democratic principles. One group supports one direction, another supports a different direction, and another group supports yet another direction; if forced to collaborate, these groups might find themselves in a paralytic deadlock. But if they are allowed to pursue their own interests, developments might occur in locales where the strongest stakeholders support a particular developmental path. This is, of course, a double-edged sword: while clean energy interests can make substantial inroads in certain locales, conventional energy interests, if strong enough in a given locale, can block progress. This fact partly explains why clean energy development in states like coal-rich Virginia is so phlegmatic.

On the other hand, when there is sufficient support from higher levels of government, polycentric groups can also find strategic alignment from top to bottom. Polycentric strategies of this type can significantly improve systematic implementation. "Nestedness," involving multiple authorities and overlapping jurisdictions, ensures that initiatives do not slip through the cracks at some level and derail the process.[55] The justification behind polycentric approaches is that conventional forms of governance such as top-down centralized control, bottom-up decentralized control, and even free-market privatization have flaws that are best resolved through consolidated strategies.[56]

Polycentric approaches can encourage plurality, promote dialogue, ensure redundancy, and enhance accountability. Involvement by multiple governance regimes ensures that different officials will review a problem from different perspectives.[57] This diversity of perspective produces a broader variety of potential solutions and facilitates experimentation. When these groups are coordinated, opportunities will arise to exchange results and observe experiences, promoting innovation and adaptive management.

Under a system of polycentrism, if one level of government, industry, or civil society ignores a problem, fails to address it, or underperforms, other governance layers can step in to fill the void. This "safety net" can ensure that wicked problems are at least addressed to a certain extent. In a best case, polycentrism can ensure that more resources are harnessed to address a particular problem.

Polycentrism can also lead to initiatives that are infectious, establishing new market tends and inspiring others to act. Consider the potential impact of a recent announcement by Xcel Energy, a large U.S. utility that serves 3.6 million customers across eight U.S. states. In December 2018, it announced a commitment to become completely carbon-free by 2050 (80 percent carbon-free by 2030).[58] Because this utility is competing for new business throughout the United States, this announcement raises the appeal of Xcel's services to carbon-conscious state governments and puts pressure on competing utilities to either keep pace or risk being shut out of certain markets.

A final benefit is that polycentrism can improve accountability, autonomy, and participation. Multiple levels of government, the thinking goes, are harder for special interests to capture and dominate. Moreover, polycentric forms of governance can create a greater sense of legitimacy and a higher level of participation since some elements are situated closer to communities and more responsive to local needs.

When applied to energy and climate change governance, polycentric forms of action can involve multiple scales (local, regional, national, and global), different mechanisms (centralized command-and-control regulations, subsidies, voluntary collaboratives, and the free market), and varied actors (government institutions, corporate and business firms, civil society, and individuals and households). In summary, polycentric efforts contribute to more effective energy governance by advancing the following:[59]

- **Equity**—Multiple representative streams mitigate the threat that some power groups get all of the benefits at little cost.
- **Inclusivity**—Multiple representative streams give everyone a voice somewhere in the system.
- **Information**—Multiple representative streams afford monitoring of programs and mechanisms at different levels, yielding richer data and higher levels of feedback and sharing of knowledge.
- **Accountability**—Stakeholders bear some of the costs of governance themselves and become accountable for their actions.

- **Organizational multiplicity**—Numerous and diverse actors are engaged, deepening coverage of implementation and enhancing monitoring systems.
- **Adaptability**—In multiple representative streams, change occurs through a gradual rollout, ensuring greater responsiveness to emergent needs.

This all sounds great in principle, but is polycentrism workable in practice? We present five case studies that demonstrate how polycentric approaches have yielded successes in energy transitions: Danish electricity policy, off-grid energy systems in Bangladesh, Chinese cookstoves, transport planning in Singapore, and building code transformation in California. Table 8.2 offers a summary of these cases, showing how they cut across sectors, countries, and periods and exhibit different manifestations of polycentrism.

Danish Electricity Policy

Denmark has relied on various policy mechanisms involving different actors to reduce GHG emissions. Policy tools have featured taxes, research subsidies, and feed-in tariffs. To list just some of the most significant achievements, Denmark transitioned from being almost 100 percent dependent on imported fuels such as oil and coal for its power plants in 1970 to becoming a net exporter of fuels and electricity. Danish wind turbine manufacturers led the world for many years in exporting wind turbine technology, at one time holding roughly one-third of the world market. As of 2018, Denmark also had the largest portfolio of wind projects integrated into its power grid of any country in the world (more than 40 percent), and some parts of the country, such as western Denmark, frequently supply more than 90 percent of electricity demand through wind turbines. As more renewable energy and cogeneration units have come to replace less efficient and more polluting conventional fossil fuel units, total CO_2 emissions associated with fuel combustion and electricity generation have plummeted. These achievements are all the more impressive when readers

TABLE 8.2 Five polycentric approaches to energy and climate governance

Topic area	Case study	Period	Description	Polycentric component	Result
Electricity supply	Danish electricity policy	1970–2001	Utilized energy and carbon taxes, R&D funds, government financing, other mechanisms to promote wind energy, energy efficiency, and combined heat and power.	Blended small-scale decentralized community control with national standards and policies.	Denmark topped the world in electricity generation from wind turbines (as a percentage of its energy mix) and wind energy manufacturing, and it leads Europe in the use of combined heat and power.
Deforestation and agriculture	Bangladesh's Grameen Shakti	1996–2009	Utilized an innovative financing scheme and market-based approach to promote solar panels, biogas plants, and improved cookstoves throughout rural Bangladesh.	Enrolled communities in projects at the household and village levels but also engaged district and national policymakers and international donors and lending firms.	Grameen Shakti operates 750 offices throughout every state in Bangladesh and has installed 250,000 solar home systems, 40,000 cookstoves, and 7,000 biogas plants, benefiting 2.5 million people.
Off-grid energy use	China's national improved cookstove program	1983–1998	Relied on a "self-building, self-managing, self-using" policy that focused on having rural citizens invent, distribute, and care for improved cookstoves.	Program was funded by the national, provincial, and local governments in conjunction with cost sharing by households.	Program installed 185 million improved cookstoves in 78% of all Chinese households, reduced energy use per capita in rural areas by 5.6%, and helped increase forest coverage in rural areas from 12% to 13.4%.

Transport	Singapore's urban transport policy	1971–2009	Restrained private automobile ownership through vehicle ownership quotas, congestion charges for roads and expressways during peak times, and proactive promotion of bus and rail mass transit.	Harnessed public–private partnerships to operate mass transit systems and worked with automobile retailers to equip vehicles with electronic road pricing devices.	Almost two-thirds of daily trips during peak hours employed mass transit, more than 95% of roads and expressways were congestion-free, and the road pricing scheme alone funneled US$138 million back into the government budget.
Sustainable buildings	California's building codes	1978–2017	The latest building codes, which went into effect on January 1, 2017, make it mandatory to provide electric vehicle (EV) charging infrastructure and EV parking spaces for new commercial and multifamily residential facilities.	These codes were adopted at the state level and indicated a proactive decision by the state to accommodate a burgeoning number of electric vehicles.	California ranks high in the overall comprehensiveness of its transportation and building energy efficiency policies. The state also provided rebates of up to US$5,000 for light duty zero emissions vehicles and plug-in hybrid electric vehicles.

Sources: The first four case studies were updated from earlier versions published by Brown and Sovacool: M. A. Brown and B. K. Sovacool, *Climate Change and Global Energy Security: Technology and Policy Options* (Cambridge, MA: MIT Press, 2011). The California building codes case study updates an earlier version (Brown, Marilyn A., and Yu Wang. *Green savings: how policies and markets drive energy efficiency: how policies and markets drive energy efficiency*. ABC-CLIO, 2015).

consider that the country is roughly the same size and population as the state of Maryland in the United States.

Denmark's successful approach to technological transition in electricity has been driven by taxes on energy usage in general, as well as on fossil fuels and carbon dioxide. These taxes created incentives for improving energy efficiency and provided government revenue for financing renewable energy research. Energy taxes also helped fund government programs to support local communities interested in using low-tech windmill designs to generate electricity. Danish regulators, working with manufacturers and interested citizens, blended a bottom-up, progressive, incremental learning model with national regulation.[60] The Danish model embraced "learning by doing" and developed systems to scaffold new learning as new needs arose. This approach paid dividends. According to researchers at the Risø National Laboratory, from 1980 to 2005, the cost per kWh of Danish wind turbines decreased 70 percent, and Danish R&D developed wind turbines that produced 180 times more electricity at 20 percent the cost.[61]

Denmark's energy transition reflects a polycentric model in action. Equity seems to play an elemental role. Denmark funded research on wind energy and cogeneration through energy taxes so that costs were spread among electricity customers for the proportions used. Denmark also attempted to partially account for the "external" costs of climate change by introducing a carbon tax, which once again held energy users accountable by the proportions used. Inclusion, especially the inclusion of rural actors, homeowners, and stakeholders, can enhance the effectiveness of energy transitions. In Denmark, guaranteed access to the electricity grid minimized barriers to market entry and prevented utilities from unjustifiably barring renewable energy projects (and other independently owned projects) on the grounds of inadequate transmission and distribution or intermittent energy flows. Danish electricity policy also highlights the importance of incentivizing communities to take ownership. In 2005, for example, only 12 percent of wind farms were utility owned; the remaining 88 percent were owned by individuals and cooperatives. In terms of organizational multiplicity, Danish electricity policy engaged with *kommunes* when setting guidelines for new energy plants. Danish policy has also been adaptive and flexible. Denmark continuously monitored market developments and progressively altered its subsidies and support for wind to control the pace of development, scaling down and eventually repealing tax credits for wind energy in the 1990s in response to the threat of excess wind capacity.[62]

Biogas and Solar PV in Bangladesh

The nonprofit organization Grameen Shakti (GS) provides financial and technical assistance for off-grid renewable energy projects in rural Bangladesh. Based on experiences from the microcredit program of its parent company, Grameen Bank, GS adopted a community-based approach to promote solar home systems, small-scale biogas plants, and improved cookstoves to reduce deforestation, fight poverty, and provide affordable energy access. Other notable elements of the strategy include a focus on conceptualizing energy supply as a small business, relying on local knowledge and entrepreneurship, utilizing community awareness campaigns, and adopting innovative payment methods that involve fertilizer and livestock in addition to cash.

The solar home systems (SHS) program, which was initiated in 1996, draws on the decentralized nature of solar photovoltaic (PV) panels to enhance energy access in rural Bangladesh. The SHS program targets those areas that have little to no access to electricity and limited opportunities to become connected to the central electricity grid within the next five to ten years. The ease of working with solar panels, their long life spans, avoidance of combustible fuel (which must be transported in), and lack of pollution make solar PV an ideal choice for remote areas. The SHS offers microcredit schemes to enable homeowners and businesses to acquire the capital needed to finance installation.

The GS biogas program promotes the use of small-scale, 3-cubic-meter biogas plants in homes and residences for providing heat for cooking. The biogas program also supports commercial development, with larger systems offering enough gas to meet the energy needs of restaurants, tea stalls, and bakeries. By utilizing biogas, these units minimize reliance on traditional forms of biomass, such as animal dung and charcoal (with their negative environmental and health impacts). The new technology also protects communities from disease by enhancing sanitation through better animal waste management.

Lastly, the Improved Cook Stoves (ICS) program distributes one-, two-, and three-mouthed clay cookstoves that cut fuel use by half and sport chimneys that create a smoke-free cooking environment. These more efficient cookstoves not only result in less fuel consumption (typically reducing fuel needs by 40 to 50 percent), but they also facilitate shorter cooking times and generate more heat.

Collectively these programs have greatly expanded access to energy services, empowered women, improved health, slowed deforestation, and helped decarbonize Bangladesh's domestic energy sector. At the end of the case study period, in 2009, GS operated offices throughout every state in Bangladesh and had installed 250,000 solar home systems, 40,000 cookstoves, and 7,000 biogas plants, benefiting 2.5 million people.[63]

The GS program, like the Danish case, illustrates polycentric diffusion strategies. In terms of equity, it spread program costs among consumers and users so that no single group had to pay an unfair or asymmetrical amount. GS linked the use of biogas units in homes and shops with fuel stock from the livestock, poultry, agriculture, and fishery industries, ensuring that adoption truly benefited a wide swath of society. Similar business linkages were made in the promotion of GS's solar panels and mobile telephones, compact fluorescent light bulbs, and light-emitting diode devices. In terms of information provision, GS held monthly meetings with all regional and division managers at its head office in Dhaka, and its staff visited SHS and biogas clients an average of once a month to check on performance and proactively address potential problems. The Grameen Technology Centers operated by GS often conducted large demonstrations of solar and biogas devices, and GS employees sometimes embarked on door-to-door visits to familiarize communities with technology and learn about their preferences and needs. The programs promoted accountability by relying on an installment-based credit system whereby the users typically paid for their own equipment. GS ensured organizational multiplicity by coordinating its activities at all scales of government in Bangladesh, integrating actors from the national, state, and local governments. GS was also flexible and adaptive, lowering the interest it charged on its microloans when it received feedback from consumers about the way it calculated energy investments.[64]

Perhaps the greatest achievement of this program was the impact it had on government policy. In 2003, the government became involved to expand the SHS program, garnering support from international aid agencies. As of May 2017, nearly "4 million solar home systems had been installed, impacting more than 12 percent of the entire Bangladeshi population."[65]

Chinese Cookstoves

Despite a trend toward urbanization in recent years, nearly 60 percent of the population in China still lived in rural areas in 2017,[66] and many households still relied on solid fuels such as wood, biomass, and coal that they burned in their homes to meet energy needs. To reduce shortages of these fuels among the rural population, reduce health costs, minimize the degradation of forest lands and agricultural soils, and promote economic growth through a positive return on energy savings, the Chinese Ministry of Agriculture initiated a National Improved Stove Program (NISP) from 1983 to 1998. Over this period, the NISP promoted the installation of 185 million improved cookstoves and increased the proportion of improved stoves from less than 1 percent of the Chinese market in 1982 to more than 80 percent by 1998. As a consequence, energy use per capita declined in rural areas at an annual rate of 5.6 percent from 1983 to 1990.[67]

The justification for embarking on a major cookstove program was related to the fact that most stoves in use at the time had efficiencies averaging 10 to 12 percent, meaning that as much as 90 percent of the energy content of the wood or charcoal used in them was wasted. In some cases, existing cookstove efficiency could be further improved by something as simple as adding a chimney or placing more insulation around the stove to retain heat. In other cases, the central goal was to replace older stoves with new stoves with efficiencies greater than 20 to 30 percent.[68]

Newer stoves significantly reduced indoor air pollution. These new models had a grate and an improved combustion chamber, and they almost always had a chimney. They utilized higher-temperature ceramics, fire-resistant material, and longer-lasting metals; had more insulation; and housed a better frame that funneled hot gases more effectively. They could cook more food at once, and many had coils around the combustion chamber to heat water while cooking. Some improved stoves were connected to radiators or space heaters so that heat could be recycled or vented to other rooms; and some stoves sent heat through pipes directly into a brick platform called a *kong* that occupants slept on at night. On rare occasions, improved stoves were "fuel flexible" and could combust coal and biomass, although doing so required a homeowner to insert a different combustion chamber for each fuel. Improved stoves were often aesthetically pleasing, with beautifully designed tile and artwork, rendering them as something to be handed down to future generations as assets.[69]

Although this program was implemented in a nation with strong central control, it exhibited a number of polycentric governance elements. The Ministry of Agriculture adopted a "self-building, self-managing, self-using" policy focused on having rural people themselves invent, distribute, and care for cookstoves. It accomplished this, in part, by setting up pilot programs in hundreds of rural provinces. One primary aim of the NISP was to train local stove builders and installers so that they could carry out most of their work autonomously. Combustion chambers and ceramic tiles were often manufactured at the village level (one hundred batches a day) instead of at huge industrial plants. Some builders specialized in stove construction, while others retained unique knowledge about how to incorporate stoves into homes. The national government was responsible only for certification and monitoring, with the remaining aspects of cookstove design, installation, and commercialization left to the marketplace—placing accountability in the hands of manufacturers and users. The program also inclusively tapped a network of service and implementing organizations spread across 38 provinces and municipal regions with more than 1,500 regional energy offices in 2,300 counties.[70]

Singapore's Transport Planning

Singapore has pursued a holistic approach to urban transport policy that involves both supply-side and demand-side targets and runs the policy instrument gambit from "carrots" to "sticks."[71] The strategy has been supported through the introduction of high-tech road pricing schemes and the provision of real-time traffic information to drivers. Supply-side components have included enhanced interconnectivity of public mass transit, large-scale investments in train and bus infrastructure, and restrictions on private vehicle ownership using quotas and taxes. Demand-side components, which aim to alter behavior in favor of public mass transit, include high-tech road pricing schemes and integrated public transit cards. One key element of Singapore's policies is that they rely on a mix of incentives and disincentives; "sticks" raise the costs of driving a private automobile through purchase taxes and road tolls, whereas "carrots" encourage public transport.[72]

These innovative mechanisms would not have been nearly as successful if it had not been for rigorous investment in public transit. Efforts began in 1973 with improvements of buses, including the forced mergers of bus companies, followed by the introduction of a state-established bus corporation in 1973, the reorganization and streamlining of bus routes, and the creation of bus lanes in major corridors.[73] Singapore spent billions of dollars building mass rapid transit (MRT) and light rapid transit rail systems, which commenced in 1987 to support the already extensive 12,600 buses and 20,000 taxis in the city. The Singaporean Ministry of Transport estimates that almost 5 million trips (about 60 percent) occur per day using mass rapid transit, light rail transit, and buses—impressive figures given that the country had a population of less than 5 million people at the end of the case study period.[74] Interestingly, one of the more influential initiatives in the promotion of public transport has been the development of covered walkways, which were expected to cover 200 kilometers by 2018, quite the feat given that the nation is only 700 square kilometers in size.[75] In Singapore, one can usually walk from home to a bus stop and then, upon alighting, onward to one's destination without having to use an umbrella.

Despite the fact that Singapore is a tightly run corporate democracy, polycentric governance principles abound in this case as well. It harnesses public-private partnerships to build and operate public transit systems and works with automobile retailers to equip vehicles with electronic road pricing devices. In terms of equity, the government provided the initial capital to build the MRT and maintains investments in rolling stock, but private companies recoup operating costs from passenger fares. Some of the money used to build and maintain this system comes from the revenues generated from the vehicle license fees and electronic road pricing. Singaporean planners seek feedback from users, and they have continually adjusted policies to make them more flexible and adaptive to commuter needs. Moreover, extensive public communications and education efforts are conducted long before the launch of any major policy changes to ensure that stakeholders have time to prepare.[76]

California Building Codes

California has pioneered aggressive building codes in conjunction with strong city ordinances and a broad inclusion of other stakeholders, such as electric utilities. Since establishing building codes in 1978, California has updated its codes approximately every three years by publishing increasingly stringent standards on the building envelope, heating and cooling equipment, lighting, water heating, and other equipment. These codes have saved more than US\$66 billion in energy costs and averted well over 250 million metric tons of GHG emissions.[77] Building codes in effect through 2016 are estimated to have reduced home energy consumption by 25 percent and reduced commercial energy consumption by 30 percent, compared to previous standards.[78] According to the California Energy Commission, in return for an investment of US\$2,700 to upgrade a single-family home, a homeowner would save US\$7,400 in maintenance and energy expenses over the life of a mortgage by adhering to 2016 standards.[79] The latest standards also encourage provisions for electric vehicle recharging infrastructure.

As one example of technological innovation incentives embedded in California's building energy codes, the 2016 code introduces requirements for adaptive lighting, with photosensors, occupancy sensors, and multilevel lighting controls. It also expands the capacity of demand response (DR) in all commercial buildings greater than 10,000 square feet by requiring new and reconstructed buildings to be designed with the capability to reduce lighting energy use to at least 15 percent below the building's maximum lighting power when utilities issue a DR signal.

In 2011, California became the first state in the nation to implement a mandatory green building code that required the state to expand from energy-efficient to sustainable buildings. The code requires that new buildings reduce water usage, divert construction waste from landfills, and use low-pollutant-emitting carpets, floors, and paints.[80] As of 2018, California ranked second on the American Council for an Energy-Efficient Economy (ACEEE) State Energy Efficiency Scorecard, with a score of 43.5 out of 50.0.[81]

At the municipal level, multiple cities in California have their own programs and policies to improve energy efficiency, particularly in the buildings sector. Los Angeles ranks fourth in the ACEEE city-level rankings of energy efficiency policies[82] and has more Energy Star buildings than any other city in the United States.[83] Improving upon the statewide initiative for encouraging electric vehicles, Los Angeles has mandated that by 2025, 80 percent of new light duty vehicles should be full-battery electric vehicles.

San Francisco ranked ninth in ACEEE's energy efficiency rankings in 2018.[84] The city's Environment Code mandates a 40 percent reduction in GHG emissions by 2025 and 80 percent by 2050. The San Francisco Building Code requires all new commercial and residential buildings to be 10 percent more efficient than the mandatory state requirements. Going one step further, Berkeley has set a goal of reducing energy usage in buildings by 33 percent by 2020 compared to 2000 levels, and by 80 percent by 2050. This goal catalyzed a 12 percent reduction in residential building energy usage between 2000 and 2012, saving more than 2 gigawatt hours each year.[85] More than 1,000 solar PV systems were installed between 2000 and 2013 in the city.[86]

The city of Oakland requires all buildings that cost more than US$3 million to meet Silver LEED status. Consequently, by 2014, 102 of the 115 largest facilities in the city had been retrofitted.[87] More than 90 percent of the city's buildings are benchmarked based on the Energy Star Portfolio Manager. Oakland's

Smart Lights program aims to improve energy efficiency for buildings that house small businesses. The city is on track to reduce its GHG emissions by 36 percent by 2020 and 83 percent by 2050 (compared to the baseline of 2005).[88]

Investor-owned utilities also play an important role in California's adoption of energy efficiency measures. For example, the Pacific Gas and Electric Company (PG&E) provides incentives to manufacturers, retailers, and customers to support the purchase and installation of efficient LED lighting products. Its partnerships with cities and across other western states illustrate how utilities can stimulate efficiency improvements across city and state boundaries.[89]

Looking to the future, California has adopted the goal of implementing zero-net energy building standards for residences by 2020 and for commercial buildings by 2030. Additionally, heating, ventilation, and air conditioning (HVAC) standards will continue to be optimized to perform most efficiently in the California climate; and by 2020, energy efficiency support programs for low-income consumers will be expanded. Strategies to ensure that these goals are met include encouraging innovative building designs, increasing customer adoption through incentives and education campaigns, increasing research funding, introducing low-cost financing options, and imposing stricter energy codes and standards.

It is estimated that by 2020 all preexisting houses in California will achieve a 40 percent decrease in energy use compared to 2008. The California renewable portfolio standard (RPS) will promote an increase of zero-energy buildings from 20 percent currently to 33 percent by 2020.[90] At the same time, the California Energy Commission has noted that additional policies and actions will be necessary to induce deeper, cost-effective energy efficiency upgrades.[91] By continuing to work across agencies and up and down scales of governance, this influential U.S. state will likely continue to provide the gold standard for energy efficiency in the United States.

CONCLUSION

Evidence suggests that holistic approaches taken together by governments, individuals, and a broad array of different stakeholders in polycentric clusters can prompt rapid, effective, low-carbon energy transitions. Danish electricity policy, off-grid systems in Bangladesh, Chinese cookstoves, transport planning in Singapore, and building codes in California offer striking examples of programs and policies that have benefited from polycentric approaches. These case studies show that communities, countries, and firms can drastically reduce GHG emissions, accelerate deployment of new energy systems, restore large tracts of degraded forestland, and greatly reduce pollution given the right mix of technologies and public policies.

Admittedly, the case studies highlight initiatives that have benefited from consistent, progressive polycentric coordination. They have been *coordinated* in the sense that policies were implemented in concert with key stakeholders from industry and civil society, *consistent* in that constant changes to program structure were kept to a minimum and were transparent when changes were necessary, and *progressive* in that they all had mechanisms for adjusting and responding to emerging challenges when necessary. However, even if such initiatives are not coordinated, it is highly likely that polycentric approaches will still perform far better in comparison to lukewarm top-down control.

The five case studies all relied on diversity to promote more sustainable policy. They depended on a suite of technologies promoted by a host of policies that influenced both technical dimensions (such as performance and price) and social measures (such as knowledge and behavior). This insight should highlight the inadequacy of attempting to address energy and climate change issues from only the supply side—building more technologies—or the demand side—altering consumer behavior. The best programs focused on aspects of both—lowering costs and improving technologies while shaping values and demand for products. This approach requires engaging with a number of stakeholders who are never easily coordinated.

These case studies also suggest that polycentric initiatives can be successful across a diverse array of political contexts. China, Singapore, Bangladesh, Denmark, and the United States represent nations that exhibit notably different political, sociocultural, and economic contexts, and yet polycentrism was successful in enabling transitions in all cases. This is not to say that polycentric governance is a panacea for all the challenges of energy governance. Powerful stakeholders can and often do capture the regulatory process and slow down transitions. Yet, in many cases, when polycentric approaches are applied, the benefits far exceed the disadvantages in regard to energy transition because any step in the right direction is a step further away from the status quo.

9

Faster, Further, Farther

Empowering the Great Energy Transition

And so, dear reader, here we arrive at the end of the line . . . give or take 10,000 words. Hopefully, if we have been effective in structuring and presenting our arguments, you will be persuaded that the trends introduced in chapter 1 present compelling—indeed, in our opinion, irrefutable—evidence that an energy transition is under way. For convenience, the list of ten trends is summarized in table 9.1.

These catalysts transcend any single dimension, sector, or scale. Some—such as declining stocks, fossil fuel price volatility, and innovations in renewable energy performance—portend technological, financial, and economic shifts. Others—such as geopolitical security concerns, evolutions in policy, or connections to development strategy—reflect an emergence of political will to change. Still others, such as the catalytic impact of climate change and other externalities, are environmental. Finally, there are other catalysts, such as opposition to nuclear power or support for renewable energy, that are underpinned by social forces. Regardless of the driving force, each trend serves as a catalyst that, even in isolation, could impact energy market dynamics

TABLE 9.1 Forces Compelling an Energy Transition

1. Growing evidence of declining fossil fuel stocks and rising prices
2. Capricious fluctuation patterns of fossil fuel prices
3. The strategic need to diversify
4. Political instability and conflict due to fossil fuels
5. Improved understanding of fossil fuel health and environmental costs
6. Sobering evidence of climate change impacts
7. The contested politics of nuclear power
8. Innovations in performance and cost within the renewable energy sector
9. The rise of government and market support for renewable energy
10. First-mover advantages amid a new energy boom

and consumer behavior. Together, these trends will create a perfect storm that will carry in its wake massive market change and will leave energy market players, especially incumbents, with the economic equivalent of storm damage, bestowing upon future generations energy infrastructure that looks very different from the rudimentary technological artifacts that form the backbone of our global energy system.

THE COMPELLING NEED FOR AN ENERGY TRANSITION

Given the perils of progressive climate change, we need to expedite an energy transition, and in doing so, we must move beyond technology development.[1] In industrialized nations, the pace of change has been too slow, and the greenhouse gas (GHG) emission reductions that have occurred have been negated by GHG emission increases in developing nations. Such phlegmatic progress underscores the critical point that the Great Energy Transition should be conceptualized as requiring the participation of a wider range of actors, including firms and consumers, as well as civil society groups, media advocates, community groups, city authorities, political parties, advisory bodies, and government ministries. Success in this transition depends not only on more comprehensive cost-benefit calculations, but also on beliefs, values, interests, resources, skills, and relations that emphasize sustainability.

This low-carbon transition will not be docile; it will be a disruptive, contested, and nonlinear process. It will be disruptive because it will threaten the economic positions and business models of some of the biggest industries on the Global Fortune 500 list (oil, cars, electric utilities, agro-food, major retailers). Resistance can be expected.[2] It will be contested because many actors still disagree about the desirability of low-carbon solutions, let alone what the solutions ought to be (e.g., onshore wind turbines, carbon capture and storage). It will be nonlinear because climate change policies and low-carbon innovations may at some junctions experience setbacks (e.g., policies in the United Kingdom, United States, and Australia) due to technological underperformance, while at other junctions they might benefit from transitionary pressures as climate change impacts worsen.[3] For these reasons, the implementation of low-carbon transitions should aim to meet a range of stakeholder needs: not only cost-effectiveness, but also social acceptance (legitimacy), political feasibility (business support), technological flexibility, and adaptive potential.[4]

Within this complex, adaptive environment, it is fair to say that developing nations have also been negligent in addressing climate change. Major emerging economies such as China, India, and Brazil are finding that the enhanced energy demand associated with robust economic growth is making it hard to stem GHG emission growth, despite awareness of the problem. Renewable energy infrastructure is burgeoning in many of these nations, but the growth in renewables has, to date, been insufficient to keep pace with the amplified demand for energy. These nations are, to invoke the lyrics of a popular U2 song, "running to stand still."

As we saw in chapter 3, the predictive accuracy of climate change science is improving, and what the science is telling us is far from rosy. Given current trends, it appears highly likely that humanity will fail to achieve the GHG emission reductions necessary to constrain global warming to the 2°C target that signatories to the Paris Agreement aspire to achieve. Indeed, there is evidence that, if trends continue, 4°C might be a more likely outcome. This prediction is troubling because 4°C portends high economic and ecological damage that can be attenuated only if we can get our collective acts together. Temperature rise projections evolve from troubling to outright alarming when one considers that many of the mainstream projections used for modeling climate change impact are highly conservative. As we saw in chapter 3, modelers have avoided the inclusion of unlikely but plausible catastrophic developments such as accelerated release of GHG stored in permafrost, escalation in the breakup of Antarctic ice shelves, and destabilization of the thermohaline circulation. In short, contrary to the toxic Kool Aid that skeptics attempts to peddle to the unwitting public, it is likely that the impacts associated with increased global warming will be more severe than mainstream models predict.

Humanity faces an *all hands on deck* challenge.[5] There is little room for placating special interests when it comes to countering what Nicholas Stern has called humanity's greatest market failure.[6] Unfortunately, if one were to throw a rock into a crowd of people anywhere on the planet, the rock would be sure to strike someone struggling with a legion of personal justifications for postponing, if not fully resisting, pressures to help mitigate climate change, adapt to its consequences, or promote more sustainable forms of energy conversion and use. Economists call this "rational inattention."[7]

As we pointed out at numerous stages in this book, energy is big business globally. Indeed, choosing from the adjectival portfolio of the current U.S. president, one could say, energy is huuuuuge. With direct economic activity generated by the energy sector estimated at 7 to 10 percent of global GDP,[8] there are vast fortunes at risk in the event of a transition. Powerful firms have billions to trillions invested in energy infrastructure that is still being depreciated, and so even if they wanted to alter course, it would take time to profitably do so. Governments have invested billions more—funds that have been largely raised through public debt that needs to be paid off. In short, the expedience of change is strongly inhibited in the short run by decisions made in the recent past when fossil fuel energy was cheap and renewable energy was struggling to compete.

With that said, it is overly simplistic to argue that the Great Energy Transition is being blocked by commercial firms and self-interested billionaires. We are all complicit in this dilemma. Many experts argue that curtailment of GHG emissions in the neighborhood of 80 percent is needed to avert the worst perils of climate change. For example, to provide a reasonable (66 percent) chance of limiting global temperature increases to below 2°C, global energy-related carbon emissions must peak by 2020 and fall by more than 70 percent in the next thirty-five years.[9] This scenario implies that we will need to triple the annual rate of energy efficiency improvement, retrofit the entire building stock, generate 95 percent of electricity from low-carbon sources by 2050, and transition almost entirely to electric cars.

Assuming that you were the only person willing to take action, consider for a moment how quickly you could enact a 70 percent reduction in energy consumption to contribute to this goal. How many automobiles does your family possess, and what types are they? How easy would it be to reduce GHG emissions associated with your transport choices, and would you be willing to do so? What is your residential energy consumption profile like? What changes would be necessary to reduce residential energy consumption by 70 percent, and how much would this cost? How about lifestyle choices? How often do you travel by airplane? Do you purchase all goods and services locally to reduce the energy consumed in transporting goods? More to the point, since much of the energy used by consumers is embedded in the products and services that are purchased, how likely is it that you could reduce your aggregate consumption by 70 percent to reduce indirect energy consumption? It is all too easy to point a critical finger at industry and fossil fuel energy firms when assigning blame for the predicament that we are in when we—fellow citizens—are the drivers of all this economic activity. Every consumption choice that we make on a daily basis goes into a global demand pool that inflates energy consumption.

Realizing that we are complicit in regard to this energy dilemma should serve as a catalyst to encourage personal behavioral change. This is not suggesting that we make changes in a manner that is either financially unfeasible or unsustainable. As the book has repeatedly demonstrated, changes can be made immediately that, over the long run, will yield a net positive economic gain. Therefore, we have elected to end this book by putting forth some viable suggestions that you could adopt in the many roles that you might play in life—citizen, consumer, peer group member, community member, worker, volunteer, advocate—in order to play a positive role in fostering a quicker transition. After examining what you can do as a citizen and potentially in your vocational capacity, we conclude with a section that speaks to policymakers who are also struggling to play a positive role in enacting change. If you are not a member of this group, you can still play a role in influencing policymaker behavior because, at the end of the day, politicians and bureaucrats earn their keep by answering to you, the voter and taxpayer.

SUGGESTIONS FOR INDIVIDUALS WHO WISH TO MAKE A DIFFERENCE

The impact that the average citizen can have on fostering an expedient transition is significant. In addition to altering your own behavior and investment patterns, you can exert pressure on family, friends, community members, and fellow citizens. Your own individual efforts might not have an earth-shattering impact, but, when undertaken in conjunction with changing market dynamics, improved education, and government persuasion, your advocacy could tip the balance. Here we describe actions that would turn an apathetic consumer, who is part of the problem, into an advocate, who is part of the solution. These actions include climate change advocacy, altered lifestyle patterns, energy efficiency upgrades, and energy transition advocacy.

Suggestion 1: Become a Climate Change Advocate

Chapter 3 attempted to paint a vista that depicts the current state of climate change science, examining scientific data and what this implies in regard to climate change impacts going forward. As careful as we have tried to be in conveying a conservative assessment of the threats associated with progressive climate change, the reader is encouraged to view all assessments (including ours) with a healthy degree of skepticism and to adopt a habit of ensuring that data and analyses regarding climate change (and anything else, for that matter) come from sources that are peer-reviewed and vetted through rigorous scientific scrutiny. In undertaking skeptical analysis, the conclusions that you reach might differ from the conclusions that we might reach. However, it would be difficult, if not impossible, to self-educate on issues such as climate change through peer-reviewed sources without coming to the conclusion that climate change represents a real and present danger, which if left unaddressed will continue to amplify extreme weather events, alter rainfall patterns, accelerate polar ice melt, and over time, elevate oceans levels across the globe. Climate change impacts, if ignored, will engender significant economic, social, and environmental costs. Since the economics of energy markets are changing in favor of renewable energy, it makes absolutely no sense whatsoever for citizens to stand on the sidelines while all of this unfolds. As climate change costs progress, your tax dollars will increasingly be channeled into adaptation and restoration projects—unless the world acts quickly and leapfrogs into a new future.

Therefore, as we highlighted in chapter 3, numerous activities can be adopted by ordinary citizens to compel politicians and friends to gain a better grasp of the science and implications underpinning climate change concerns. Rather than focusing on short-term self-gains, we have a moral imperative to include the costs that future generations will incur in any cost-benefit analysis. This moral imperative is operationalized by taking actions to defend the interests of future generations. Many would agree that we must also protect the nonhuman species on our planet that are unable to protest the threats to their existence.

Table 9.2 summarizes a few of the actions that you could take as a climate advocate to ensure that fellow citizens and politicians are making decisions based on sound scientific judgment, rather than allowing skeptical journalists to perpetuate mass societal ignorance of what is actually transpiring.

Suggestion 2: Alter Your Lifestyle Patterns

It should be apparent that our consumption habits are not constructed solely within the sanctuary of our inner wills. Indeed, after reading chapter 7, one would be excused for wondering if any of the consumption decisions we make are actually our own. As we saw, our consumption decisions are made within a complex sociotechnical system in which affluence, peer influence, education, past experience, and, yes, corporate marketing efforts combine to shape lifestyle patterns and consumption behavior. Not you, you say?

TABLE 9.2 Examples of Climate Change Advocacy You Could Undertake

- Write to your political representatives to let them know that you are supportive of climate change mitigation policies.
- Form community groups to help disseminate knowledge on climate change and encourage the adoption of lifestyle choices which will not exacerbate greenhouse gas emissions.
- During elections, ensure that you vote for politicians who support climate change mitigation and, if possible, make the effort to lend support to the campaigns of such politicians.
- Embrace small initiatives such as purchasing carbon offsets, supporting tree planting programs or encouraging energy efficiency. Moreover, enlist the participation of others.
- Help to organize cooperative investment efforts in support of GHG mitigation when government efforts are failing or nonexistent.
- Hold the media accountable by lodging complaints or writing letters to the editors when encountering irresponsible reporting.
- Develop an online presence to counter misinformation campaigns or to disseminate information to others.

Consider first how your economic status influences when, where, how, and why you do things. On vacations, what do you do? Do you undertake activities that consume energy? Do you travel afar, or do you remain in your community? If you earned half the income that you currently earn, would your vacation patterns change? How about if you earned double your income? Decisions made regarding how one commutes to work, the types of appliances one purchases for the home, and the choices one makes in regard to entertainment are all impacted by economic status. Now ask yourself one final question: has any increase in economic well-being that you've experienced resulted in a corresponding increase in happiness due to the purchases you can now make? If you are uncertain of the answer, simplifying your lifestyle by reducing consumption might add value that extends beyond the energy savings you would capture.

Consider now the role of peer influence. Do the things that you own serve purely utilitarian purposes, or are your purchase decisions made after benchmarking with the purchases of others in your group of peers? If your behavior is driven only by the former, you might want to reward yourself by making yourself a nice cup of coffee with your new capsule coffeemaker that your friend suggested you purchase. You might even want to alert others to your enlightened purchase by making a post on your Facebook account. Oops, now your independence has been shattered.

We consumers are complex entities. Our lifestyle patterns and consumption behaviors are impacted by a significant number of variables. Indeed, it is hard enough self-diagnosing our own behavior, let alone endeavoring to try to understand the behavior of others. We are influenced by peers, existing lifestyle patterns that constrain our ability to alter behavior, self-image and ego, varied and often conflicting motivations, irrational decision patterns, insufficient information, and the fickle hand of time, which alters our tastes and consumption patterns. If we wish to change and become part of the solution, we first need to try to understand our motives and reflect on the factors that drive our lifestyle patterns and consumption behavior.

Self-reflection and self-awareness are difficult. They take time and a degree of honest reflection and analysis that most of us find uncomfortable. So why undergo a process of painful self-reflection for the sake of altering our energy consumption behavior? After all, in the grand scheme of things, chances are that any lifestyle changes that we make will be offset by the growing avarice of others.

We contend that justification for change lies in the quest for happiness. The conceptual string that ties virtually all self-improvement literature is based on the premise that personal development starts with the capacity to self-reflect. This reflection allows us to establish a starting point in identifying strategies for self-improvement.[10] Seen from this context, evaluating one's approach to energy consumption requires an evaluation of underlying lifestyle patterns. When considering lifestyle changes to reduce energy usage, one is compelled to prioritize activities by considering whether the activity or behavior adds personal value. Does one gain more value from a nightly habit of television watching, or would there be more value derived from social interactions or spending more time with one's children? Does the value encapsulated in the convenience of driving down to the corner store exceed the value encapsulated in taking the time to walk there? For each person, answering questions such as these is clearly subject to individual interpretation. However, going by the plethora of self-help books that are sold around the world that stress enhanced mindfulness and simplification of one's lifestyle patterns, it is safe to say that for many individuals, reducing energy-intensive activities requires a lifestyle shift to activities that will also enhance one's sense of self-fulfillment and satisfaction. Kumbaya, friend.

Suggestion 3: Upgrade Energy Efficiency

If one were to highlight one insight on energy efficiency initiatives, the insight would be that a great many energy efficiency upgrades yield positive economic returns. So why doesn't the average person reside in a home that has extracted all of the positive returns from energy efficiency upgrades? Part of the answer centers on agency issues that are difficult to overcome, such as the reluctance of a renter to invest in a more energy-efficient furnace for the rental property. However, for many consumers, knowledge gaps and apathy play central roles in inhibiting action. Let's explore both because these problems are inherently rectifiable.

In regard to knowledge gaps, even householders who make an effort to explore technologies that can improve energy efficiency in profitable ways find it difficult to keep up with the glut of emerging innovations that permeate the energy efficiency technology sector. However, help is but a mouse click away. *Better Homes and Gardens* has a webpage that outlines 24 energy-saving tips for the home,[11] while the North American energy firm Direct Energy betters this with a webpage that offers 25 tips.[12] The United Kingdom's Green Age bests these efforts with a webpage that offers 100 tips for the home.[13] Not to be outdone, U.S. energy retailer Alliant Energy hosts a webpage that lists 101 ways to save energy,[14] while the United Kingdom's Ovo Energy takes the crown by offering 120 energy-saving tips.[15] Many of the tips found on these sites focus on conservation, so the savings require no upfront investment.

There are also sites for techies. There are many blog sites such as *Make Use Of*, which hosts a blog that introduces 7 energy-saving technologies.[16] A site called *moneycrashers* helps the visitor to quantify the costs of 10 common household energy-saving technologies.[17] Australia's Powershop hosts a webpage that delves deeper into household technology, providing an overview to 23 gadgets to improve energy efficiency.[18] These are far from the definitive sites on energy technology for the home, but they do represent starting points that can be accessed with a smart phone and a smart use of time.

For larger users of energy, there are likely numerous energy consulting firms near your home or workplace that offer more sophisticated audits of your energy consumption patterns. These range from major firms that work with major industry, such as GE Energy Consulting,[19] or major utilities, such as EVA in the United States,[20] to household energy consultants.[21]

The other influence that tends to prevent individuals from extracting positive returns from energy technology upgrades is apathy. For many energy-saving investments, the payback period is longer and not as lucrative as one might hope. Consequently, the incentive to undertake such an investment is low. Care for an example? Stop reading for a moment and take a look around you. Look for any electrical devices that are plugged into a power source. Are any of them in standby mode? If so, you are observing an example of power being wasted. Do you have any light bulbs that could be replaced with more efficient, low-wattage alternatives? If so, you have been paying more for your lighting than you should. Do you have older appliances that might consume energy in a less effective manner than new models do? If so, once again you are paying more for your energy than you should. For each of these upgrades, the upfront switching costs, the small amount of savings to be harvested, and the inconvenience of replacing these products conspire to engender a level of apathy that inhibits change. So if you want to make a difference, you need to shake yourself from this paralysis. Take the list of energy efficiency initiatives from your Internet search and conduct a brief energy audit in your home. Identify where the savings can be had and make the switch.

Suggestion 4: Become an Energy Transition Advocate

In chapter 7, we wrote about how the evolution of energy technology has physically distanced the consumer from the energy production process. Most of us, at least in developed nations, no longer participate actively in the procurement, transport, or management of the energy supply that fuels our homes and workplaces. Instead, we get our energy by flipping switches or by stopping off at the local gas station, rolling down our windows and yelling *fill 'er up*. Consequently, there is a spirit of "out of sight, out of mind" when it comes to energy procurement. This also holds true for energy efficiency initiatives. When was the last time you sat around with a bunch of friends discussing how to conserve energy or save on energy costs by investing in new technologies?

If it is true that there are many initiatives that people can adopt to save money, improve energy efficiency, and use fewer fossil fuels, the main challenge should be to ensure that everyone knows what these options are. Each of us could have an impact on energy consumption, not just by altering our own energy efficiency habits but also by alerting others to the savings. In short, if you want to make a real difference, you could potentially have more impact by being an energy efficiency or renewable energy advocate than by simply enacting change within your own homes. The impact that you could have by adopting more energy-efficient practices within your home has limits. However, if you advocate the adoption of similar practices to friends, neighbors, and colleagues, the impact that you could have is limited only by the number of people in your network. Thus, if you purchased an energy-efficient heater that saved you $100 each month in heating costs, the impact that you would have as a champion of climate change mitigation practices would be limited to the energy that you saved. But if you told ten friends about the savings you enjoyed and they all purchased the same product, not only would you have saved these friends some money, but you would also have increased your impact as a champion of climate change mitigation practices by encouraging energy savings that were ten times greater than what you saved by yourself.

Some individuals have already taking this prescription to the next level by reconceptualizing energy advocacy either as a new vocation or as a social venture. The sites listed in the previous section to inform and help people make better energy decisions at home were all started by someone, so why not you? Conduct an Internet search of organizations in your community that provide energy advice or consultancy services. Is there a market niche? Perhaps begin dabbling in this sector by starting a blog. As you gain experience, you can become more involved. In the event that a local community service does not exist, consider providing such a service by compiling insights available on the Internet and offering these insights to fellow citizens, either by using an Internet platform, establishing a consulting service (for-profit or not-for-profit), or working through a local community volunteer group.

How to Win Friends and Save Money

The remarkable thing about becoming an energy transition advocate is that it pays off in social dividends. If you want to create goodwill among members of your community, showing people how to save money is a very good strategy. In other words, having an impact as an energy advocate can increase the quality of your life by enhancing the quality of social interactions, much in the same way that altering your lifestyle patterns to reduce the energy consumed can enhance your sense of personal well-being and happiness.

SUGGESTIONS FOR ORGANIZATIONAL LEADERS WHO WISH TO MAKE A DIFFERENCE

This section treats the term *leadership* in the broadest possible sense of the word. Whether you are an entry-level employee or the chief executive officer of an organization, you have the capacity to proactively demonstrate leadership by thinking strategically and making decisions that will benefit the organization. This section should be relevant to people in any vocational capacity, although clearly the ability to adopt some of these initiatives increases as one moves up the organizational hierarchy.

Since the 1990s, people interested in the nexus between corporate strategy and environmental governance have been treated to some insightful literary works. One of the earlier works was that of Michael Porter and Claas Van der Linde, who put forward a seemingly obvious but frequently unrecognized premise that pollution and waste are evidence of suboptimal resource use. Gas pollutants are, at their core, evidence of incomplete combustion of a resource. Liquid pollutants are, at their core, evidence of potential resources that are simply being washed away with water. Waste represents an inefficient use of resources.[22] In the late 1990s, a book called *Natural Capitalism* was released that provided case study examples showing how innovative firms reconceptualized business operations to minimize pollution and waste and in the process saved millions of dollars.[23] One of the authors of this book (Valentine) investigated the numerous ways that organizations could leverage enhanced environmental governance to achieve higher levels of organizational performance; he concluded that strategic initiatives in regard to environmental governance could reduce operating costs, enhance product quality, enable premium pricing, attract and retain loyal customers, attract and retain high-performing employees, and minimize the cost of capital.[24] In short, all signs point to a glut of benefits to be achieved through resource conservation.

Given that energy costs represent sizable outlays for most firms, it follows that these organizational benefits apply to energy conservation as well. Initiatives in this vein range from low-hanging cherries (identifying simple ways to conserve on energy) to cost-saving technical upgrades to marketing campaigns designed to highlight a firm's competitive prowess in regard to carbon footprint reduction.[25] While some energy efficiency initiatives require high-level investment approval or specialized knowledge to adopt, other initiatives can be undertaken by virtually anyone in a given firm.

Table 9.3 lists organizational activities that could be undertaken to improve energy efficiency. They are far from exhaustive, but they hopefully show that there are solutions that any employee can champion. One might find that in the course of championing such activities, one is able to distinguish oneself as a take-charge leader, thereby enhancing prospects for career promotion.

Applying the spirit of scientific skepticism that we advocated for vetting climate change science, one might want to question if it is truly possible that energy conservation or energy efficiency initiatives can profit a firm. The evidence suggests that the answer is a resounding "yes." Consider the following success stories. One of the early adopters of energy efficiency improvements is DuPont, which announced an intention to reduce GHG emissions by 65 percent between 2000 and 2010. As early as 2007, DuPont discovered that it was saving US$2.2 billion a year through these efforts.[26]

TABLE 9.3 Examples of energy efficiency initiatives that could be adopted by organizations

Initiatives that anyone could lead:

- Take the initiative to search for ways to conserve energy by altering your own work behaviors or proposing energy saving initiatives.
- Form working groups to investigate ways to conserve on energy or improve energy efficiency.

Initiatives that section managers could lead:

- Design incentives to encourage employees to alter practices or innovate in order to conserve on energy or improve energy efficiency.
- Commission energy audits to examine sub-optimal energy use and to highlight profitable initiatives.

Initiatives that senior management could lead:

- Form high-level strategic working groups to investigate ways to alter organizational practices to conserve on energy or improve energy efficiency.
- Establish benchmarking metrics and performance standards in order to encourage progressive improvement in energy conservation or improved efficiency.
- Undertake comprehensive cost-benefit analyses of all energy consuming activities in order to identify profitable areas for conserving energy or for enhancing energy efficiency through technical upgrading.

Ten years ago, General Electric introduced a program entitled the "Ecomagination Treasure Hunt." The program incentivized GE employees to identify initiatives to reduce energy usage, which in the end resulted in US$150 million in energy savings and identified ways to offset over 650,000 million tons of CO_2 emissions.[27] In 2006, Tesco began examining options for improving energy efficiency, and by 2010 the firm was saving over US$320 million a year thanks to these initiatives. Similarly, in 2008 AT&T announced a portfolio of 8,700 energy projects to reduce costs, and by 2010 the firm was saving US$86 million a year in energy costs.[28] In 2014, United Continental reported annual savings of US$343 million thanks to initiatives to reduce fuel use.[29] Similarly, in the same year, Arcelor Mittal, the global steel company, saved nearly US$200 million thanks to energy-saving programs.[30]

Promisingly, some firms have adopted energy improvement initiatives as part of a broader set of corporate social responsibility commitments, viewing these commitments to be the cost of doing business, only to discover that there were no costs. For example, in 2007, Marks & Spencer announced a sustainability plan that it envisaged would cost £200 million over five years, but by 2011–2012, the firm discovered that it had saved £105 million through these initiatives. Indeed, a study conducted by Australia's Climate Works posits that 70 percent of companies could profit from improved energy efficiency. More specifically, the study concluded that a third of the companies examined could increase overall profits by 5 percent per year.[31]

This is not to say that all energy-saving initiatives will be profitable for a firm, but enough firms have reported success in this regard to warrant some consideration from a strategic level. At the operations level, anyone in virtually any workplace should be able to identify profitable energy-saving initiatives. Consider your own workplace. Is there evidence of wasted energy use—equipment, lights, or air conditioning or heating left on when not in use? Could some rooms benefit from the installation of motion sensors to manage lighting and heating and cooling? Is there old equipment that is currently in use that would justify replacement through energy savings? None of this is rocket science. It merely takes a bit of thought and some initiative—two attributes that everyone has.

SUGGESTIONS FOR POLICYMAKERS WHO WISH TO MAKE A DIFFERENCE

If you're a planner, regulator, or public policymaker who wishes to make a difference and you've reached this stage in the book, then you've already succeeded in implementing the first of our recommendations—informing yourself of the forces that deter energy transitions and seeking out strategies for overcoming these barriers to change. There are some specific insights that were scattered through this book, which merit highlighting in this section because for policymakers, being motivated to act as a champion for change does not guarantee success. A concerted effort must be made to offset barriers. Policy has an indispensable role to play in this quest.

Suggestion 1: Engender Public Awareness

We live in a world where two developments have come together to complicate the ability of the average citizen to make responsible decisions when it
comes to governance of any policy area. The first development is the proliferation of information. Not only is there more of it, but it is also difficult
to properly vet the sources and verify the data and underlying assumptions.
The second development stems from the amplification of the financial commitments made by many firms. The globalization of markets has produced
powerful multinational firms that have a lot at stake when it comes to market change. As a result, these firms are willing to spend enormous sums to
defend their markets. The combination of these two developments engenders an information environment that has recently been dubbed "fake
news." Such news is not always "fake" per se, but it has been filtered through
analytical lenses that favor vested interests. The end result is that the average citizen no longer knows what to believe.

It is for these reasons that one of the instrumental roles of a modern democratic government is to provide channels for accessing information that is
as unbiased as possible, or at least to ensure that information conduits are
explicit in regard to their biases. This is not to say that governments should
nationalize media sources, but rather that government websites should
attempt to direct citizens to sources where they can obtain information that
allows them to make responsible decisions as voters. Thomas Jefferson used
to say that "a democratic society depends upon an informed and educated
citizenry," and this necessitated educating the people "even against their
will."[32] Energy information and education campaigns to educate the public
can include grade-school classes on energy and the environment; public
demonstrations and tours of clean power facilities; mandatory disclosure
of electricity usage for the construction of new buildings and the renting
and leasing of existing ones; free energy audits and training sessions for
industrial, commercial, and residential electricity customers; improved
labeling, rating, and certification programs for appliances and electrici-
ty-using devices; and the creation of websites, free books, indexing services,
and libraries to help consumers gather and process information in order to
make more informed choices about their electricity use.

Government-led educational initiatives have at least two main purposes. The first is to direct interested parties to information sources where operational problems can be solved. The second is to dispel misunderstandings in order to engender public support.

As an illustration of government-led efforts to connect solutions to problems, in the United States, the federal government manages over twenty information programs and assessment centers that focus on educating industrial customers who wish to improve energy efficiency. The Department of Energy (DOE)'s State Energy Program, for instance, provides technical expertise to help industrial plant managers identify opportunities for energy efficiency improvements. Its Climate Vision program and the EPA's National Pollution Prevention Vendor Database have more than 1,200 listings of pollution prevention products and services. The EPA's VendInfo database helps industrial clients find energy service companies willing to install energy-efficient products. The Department of Commerce runs a Manufacturing Extension Partnership to help update plant managers and provide specialized advice on improving energy efficiency. Narrowing the focus further, the DOE's Golden Carrot program offers targeted information for school lunch programs and refrigerator manufacturers in order to enhance energy efficiency in school cafeterias.[33]

Information programs, however, must be carefully tailored. Information is less likely to be used if the source requires effort to navigate or does not provide the information in a timely manner (when needed). Householders consume energy from different technologies and in different kinds of buildings, and the usage varies in terms of income, housing tenure, and individual needs. To avoid adding to the information overload, "general" or "generic" distribution strategies must be avoided.[34]

In regard to dispelling misunderstandings, government websites can play vital roles in ensuring that all stakeholders are working from the same scientifically grounded playbook. Sites that provide a lay understanding of emerging climate change science and that allow more sophisticated users to delve deeper into subtopics should be commonplace both at the national and subnational levels. Similarly, websites that update stakeholders on the state of affairs regarding energy technologies are also important because—as we saw in chapter 1—much of the opposition to energy transition is based on arguments that have been invalidated because technological change has altered market dynamics.

Two of the most prominent and misguided critiques underpinning a renewable energy transition are that the transition will engender costs that will damage economic development prospects and that the infusion of stochastic renewable energy flows into a grid will destabilize the electricity supply, curtailing supply. As discussed in chapter 1, the economics of energy have changed significantly over the past decade. In many cases, some of the renewable energy technologies are now more economically attractive than fossil fuel technologies. This fact does not mean that switching from one technological platform to another would be cost-free; however, in the medium to long term, there is evidence that such a switch would actually be a positive investment for most nations. Many citizens do not realize this truth. In regard to concerns over destabilization of the electricity supply, there are examples around the world where renewable energy contributions amounting to as much as 30 to 40 percent of the energy mix have been safely accommodated through the spare capacity that already exists in established grids. Beyond 30 to 40 percent contribution levels, new energy storage technologies are being introduced that enable the safe storage of surplus energy at affordable rates. For example, Tesla has recently completed a battery storage facility in Adelaide that allows the municipality to incorporate higher levels of renewable energy without onerous costs while ensuring that resilience of the electricity grid is maintained.[35] Government websites that dispel such misunderstandings by connecting citizens to emerging science and engineering insights can go a long way to engendering public support for change.

One of the key caveats associated with government information provision is an acknowledgment that information is rarely neutral. Even "facts" (or at least interpretations of facts) can be contentious. Therefore, we recommend that governments that strive to provide citizens with conduits to become better informed should also endeavor to couple such efforts with access to open forums for scientific challenge and public discussion. At the end of the day, a government that wishes to enable participatory democracy needs to ensure that citizens are well informed and have access to policymakers to convey their perspectives.

Suggestion 2: Establish Benchmarks to Measure Progress

Carbon footprint analyses have become more sophisticated and are now standard components of city planning, enabling low-carbon strategies to be identified and managed.[36] More comprehensive indexes have emerged in recent years, including the genuine progress indicator (GPI),[37] which tracks twenty-six economic, environmental, and social parameters, including factors like resource depletion, pollution, income inequality, and cost of commuting.[38] The environmental performance index (EPI), a Yale-based initiative, includes twenty ecosystem and human health indicators in 180 countries.[39] Finally, the Chinese government has adopted a set of twenty-six key performance indicators (KPIs) to assess the sustainability performance of China's eco-cities.[40]

Each of these benchmarking programs provides insight into approaches that can be modified to suit community, municipality, regional, or national energy governance programs. As the adage admonishes, one cannot manage what one cannot measure.

Suggestion 3: Establish Standards and Infrastructure to Empower Innovators

For integrated deployment of smart grids, development of standards at the subnational, national, and global levels will be particularly important for optimizing technological synergies. Policymakers who hope to support developments of this type are well advised to designate lead agencies to coordinate efforts at various levels of government to ensure that there is standardization and cooperation at all levels.

Standardization within the electric power industry is particularly important because there are a number of new revenue streams that can be adopted to optimize the benefits from transitioning to a low-carbon economy. A policy framework that attracts private capital investment and encourages public-private partnerships will go a long way to help reduce the public costs associated with a national energy transition. Encouraging competing business models to enhance penetration of renewable generation technologies and to improve energy efficiency in an inclusive manner would help to improve stakeholder participation and engender community support. This all suggests that the traditional electricity grid model, predicated upon large, nationally owned utilities, needs to be replaced by a model that encourages competition in electricity generation and energy efficiency consulting services. Due to the high costs of grid infrastructure, evidence suggests that a model that encourages public management of the grid but private contributions from electricity generation firms will be the most cost-effective.[41] When competition at the electricity generation level is encouraged, cooperative investment and community-generated smart-grid developments will be able to join the party. This suggests that transitional infrastructures must be deployed to support a dynamically evolving system. The infrastructure must be flexible, open, and interoperable, ensuring that companies can benefit from access and customers and third parties can benefit from the data that they generate.

The case studies from chapter 4 also highlighted some caveats to smart-grid development that policymakers should consider. As we saw from the experience of the New York Public Service Commission, energy efficiency programs can be net positive for a community but at the same time give rise to higher costs for certain segments of the community. Accordingly, policymakers need to be aware that transfer payments or other strategies might need to be adopted to avoid disadvantaging less affluent consumers. In the Italian case study involving Enel, we saw how efforts to integrate decentralized generation benefited significantly from smaller-scale pilot projects that helped engineers learn through doing. The importance of piloting projects before mass market rollouts was also highlighted in the South Korean case study, which provided a good example of how larger regions could approach technological transition in logistically manageable chunks.

In addition to discussing government support for infrastructure buildout, chapter 4 highlighted the benefits associated with extracting better productivity from existing assets.[42] Improving the productivity of existing assets reduces the total amount of investment necessary for supporting energy transitions. This fact suggests that efforts need to be put into place to encourage a sharing economy. Car-sharing programs, bike-sharing programs, subletting of public facilities, and programs designed to make use of energy during nonpeak hours all hold much promise for improving energy productivity. However, all of these new businesses need to have market development support, and in some cases infrastructure or startup support, to seed investment. For policymakers, there is strong justification for supporting such initiatives because a sharing economy tends to be localized, retaining the profits in communities.[43]

Suggestion 4: Catalyze Investment from Multiple Stakeholders

From even a cursory reading of chapter 5, it should be clear that an expedient transition to low-carbon energy systems is going to be very expensive. Indeed, for many nations, financing such an outlay with public funds over a short period of time would be prohibitively expensive. Innovative approaches to financing are therefore needed if policymakers are to foster expedient transitions.

Governments will be hard-pressed to avoid funding some elements of the transition. However, with some creativity, the public expenses might be far less than one might anticipate. For example, electricity supply management costs can be reduced in many nations. There is considerable evidence that a publicly managed electricity grid is in the public interest because security of the grid is a vital component of national security and is therefore a public good.[44] Maintaining a national grid is also cheaper if economies of scale can be tapped, which are a characteristic of many large state-owned utilities. So governments are usually on the hook for the grid. However, it should be apparent after reading through the content in chapter 4 on technological infrastructure that private, cooperative, and community financing might be able to help establish decentralized supply nodes that can energize the grid, thereby reducing the costs of subnational connectivity. Similarly, setting standards and regulating electric vehicle charging stations might be unavoidable government costs, but because of the profit component of providing such access in public (and private) venues, there is no reason why the private sector cannot provide charging stations.

Governments will also need to play a role in supporting research and development (R&D) if we are to continue to develop commercially transformative technologies. However, as the Mission Innovation and the Breakthrough Energy Ventures initiatives outlined in chapter 5 illustrated, some opportunities for cross-boundary cooperation and collaboration with industry can help to reduce public research costs while simultaneously connecting with the stakeholders who will eventually be responsible for commercializing any innovations.

After markets are liberalized to allow prosumers, entrepreneurs, cooperatives, communities, and other private enterprises to play a role in supporting infrastructure buildout, and after inclusive R&D networks are developed that bring together public and private funding, there will still be sizable investment requirements that governments will have to undertake on behalf of their citizens.[45] Accordingly, it should be clear after reading through chapter 5 that carbon pricing of some sort needs to be a driver for funding the transition. Regional, national, and subnational carbon pricing schemes already cover 22 percent of global CO_2 emissions, indicating that carbon pricing is on a trajectory that will make it the norm a decade from now. Nations that currently do not have carbon pricing schemes are missing out on golden opportunities to seed energy transitions during a sensitive developmental era when pioneering firms and nations stand to gain a competitive leg up in an industrial sector that will likely be the largest in the world in coming decades.

It's important to note, though, that carbon taxes should be seen as just the first stage in a program to expedite an energy transition. The second stage is for these tax revenues to be utilized in an effective way to facilitate capacity expansion and enhanced R&D. Failure to do so will simply draw out the transition.[46]

Policymakers also need to be aware that carbon taxes can give rise to unintended consequences. Carbon taxes alter the fortunes of firms and can lead to transitionary unemployment and economic woe in certain communities if plans are not put in place to ensure that local fossil fuel enterprises that are damaged by these taxes are replaced by emergent clean-tech businesses. Policymakers should also be aware that carbon taxes can penalize poorer consumers, who use a larger share of household earnings to buy electricity and gasoline.[47] The point is that carbon taxes need to be strategically designed and implemented.

Suggestion 5: Get the Policy Mix Right

An important initiative that policymakers can undertake is to commission a strategic review of the policy environment as it pertains to a given energy market when there are aspirations to invoke change. As chapter 6 pointed out, the process of developing a policy portfolio to enact change must begin by understanding the nature of the problem. The social, technological, economic, environmental, legal, and political characteristics that underpin any national or subnational energy market must be thoroughly studied, and barriers to change must be comprehensively documented and understood before strategies can be created to actuate change.

In the previous two sections, we alluded to some important policy prescriptions that policymakers should consider as part of the strategy. In order to promote technological innovation and transformation, policymakers need to consider R&D subsidies, research grants, and tax rebates to encourage progressive innovation. Electricity markets must be structured in ways that allow multiple stakeholders to play commercial roles in this transition. Monopolies (aside from grid management services) need to be broken up, and stakeholders at various levels, from householders to cooperatives to private firms, need to have guaranteed access to electricity grids. Standards and cybersecurity safeguards must be maintained to preserve the public interest, but entrepreneurs and innovators need to be supported through public agencies that have the flexibility to promptly take down artificial barriers that thwart progress.

In order to muster the financing necessary to support a widespread energy transition, tax regimes need to be altered to entice desirable investment and discourage counterproductive investment behavior. Subsidies for technological upgrading and for enhancing energy efficiency should be considered alongside taxes and other disincentives applied to carbon-intensive energy activities. Policies must be put in place to support entrepreneurial forays at the household, cooperative, community, and corporate levels. Nodality instruments designed to educate citizens and businesses need to be structured in a way to entice participation, and government agencies need to be established to provide consultants for services to help stakeholders engage financially.

As chapter 6 pointed out, in many ways policymaking is partly a craft as well as a science. Globally, there are numerous case studies on how to effectively structure regulations, subsidies, taxes, carbon trading programs, and other policy tools. However, as we also explained in chapter 6, creating energy policy is a contextually infused process. Policies that have been successful in some markets might fail in epic fashion in other markets. Therefore, policymakers who are intent on creating policy portfolios to drive the transition need to also have in place monitoring systems to evaluate effectiveness. Subsequently, they must instill flexibility within the bureaucracy to allow a change in course, should progress not be as envisioned.

It also merits noting, as we did in chapter 7, that there is a high degree of consumer apathy associated with the energy predicament in which we are currently embroiled. For each policy environment, policymakers need to make the effort to understand consumer behavior at a far deeper level than is currently the norm. Educating the public on how to play a role in fostering an energy transition might not be enough to break free from the path-dependent behavior that consumers around the world have adopted, behavior that leads them to exhibit suboptimal behavior when it comes to energy consumption.

Certain technologies may need to be either prohibited or heavily taxed to discourage continued usage. In chapter 7, we highlighted some of the factors that inhibit change at the consumer level. We noted that consumer behavior is on one level heavily influenced by social constructs, by peer pressure, and by media influence. None of these factors tend to be characterized as forces that support energy conservation. Consumers are also often either misinformed or uninformed. In regard to energy, there are some initiatives that consumers could adopt to actually save money, but they don't do so because they are unaware of these opportunities. On another level, though, consumer misinformation or ignorance is also partly a result of disinterest. People simply have other things to do, and conserving energy or improving energy consumption is usually not near the top of one's "to do" list. This is where government policy has to come in. If policy can help to inform consumers, alter behavior by minimizing the costs of adopting advanced technologies over others, and promote technological upgrading by minimizing switching costs, then massive changes are possible.

Suggestion 6: Harness Social Networks to Invoke Change

The final area where policymakers could have a major influence in supporting energy transitions is by doing something that governments at most levels are quite proficient at—coordinating networks. In chapter 8, we talked about the allure of polycentrism: coalitions of citizens, interest groups, corporations, and multiple levels of governance working in a loosely affiliated manner to achieve a common goal. In a democracy, when it comes to the creation of policy to govern an area where commercial activities take place, there are bound to be competing interests. It is unrealistic to expect all stakeholders to be willing to paddle in the same direction when expediting an energy transition. Indeed, while some are paddling, others will be attempting to capsize the boat. But the conflicts that exist in policy circles in democratic nations need not result in paralysis.

For those who are most concerned about the pace of climate change, the speed of energy transition will never be sufficient. However, the pace of transition can be accelerated by improving the coordination through which networks strive to engender change. Today, connectivity is at the greatest level that it has ever been. New initiatives can be shared between communities in the blink of an eye. People who need to learn how to do something can gain a full education through a few mouse clicks. Entrepreneurs can be connected to financiers, companies can form alliances, and support can be mobilized in rapid-fire fashion thanks to the Internet age.

Governments already have gained a degree of competency in linking networks together. Click on the municipal website for virtually any city and you encounter vivid examples of this interconnectivity. For example, Stockholm's website (visitstockholm.com) supports tourism in the Swedish city by providing information on restaurants and events and links to numerous public and private websites that might be of use to visitors. Like other similar tourism websites, it is a clear example of the capacity of a government to serve a comprehensive networking function by connecting site visitors to public and private organizations. In regard to promoting energy savings, the New York Energy Research and Development Authority (NYSERDA) established a website that promotes energy efficiency and the use of renewable energy sources.[48] The site is designed to provide customized services and links for business and industry, communities and governments, residents and homeowners, partners and investors, and researchers and policymakers. In addition to outlining programs and services offered by NYSERDA, the website boasts a portal that allows visitors to find a contractor in their area who can help advise on energy-saving initiatives. This is a highly relevant and replicable example of how one government agency can enhance policy effectiveness by promoting access to multiple stakeholders, thereby providing more targeted support for those hoping to play a role in invoking such a transition.[49]

The reader is reminded to revisit chapter 8 and the case studies involving Danish electricity policy, off-grid systems in Bangladesh, Chinese cookstoves, transport planning in Singapore, and building codes in California for examples of programs and policies that have benefited from polycentric approaches. For policymakers who are working within organizations that are particularly phlegmatic in regard to supporting energy transitions, these types of polycentric approaches represent work-arounds that can be adopted to make an impact despite the existence of obstructive policy regimes. In short, even policymakers who are working within dysfunctional agencies can play a role and garner inspiration by searching out collaborative approaches for making a difference.

RUMBLE IN THE JUNGLE REDUX

On October 30, 1974, undefeated world heavyweight boxing champion George Foreman squared off against former heavyweight champion Muhammad Ali in Zaire in a boxing match that would come to be known as "The Rumble in the Jungle." Foreman at the time was a youthful twenty-five years old and had a boxing record of forty wins and no defeats, with thirty-seven knockouts. He was big and strong and was dominating the sport. Ali was then thirty-two years old, and many believed him to be in the waning days of his career. Indeed, there was little indication that Ali stood a chance against this powerhouse of a boxer.

But Ali was always one to surprise. In the prelude to the fight, he arrived before Foreman and endeared himself to the fans in Zaire. He taunted Foreman mercilessly, and by the time the day of the fight arrived, Foreman was incensed. From the beginning of the fight, Foreman was a raging bull, doggedly pursuing a ducking-and-dodging Ali around the ring. Ali would periodically interrupt his evasive tactics to throw sharp right-hand leads at Foreman that seemed to further incense his powerful challenger. Then in the second round, Ali introduced a new tactic—the "rope a dope." Standing with his back to the ropes, he covered his head and body and used the elasticity of the ropes to help him evade damage and counterpunch effectively as Foreman wildly rained down ineffective punches. As the rounds progressed, it became apparent that instead of Ali being mortally injured by Foreman's onslaught, his enraged opponent's energy was beginning to wane. In clinches, Ali would chide Foreman, "They told me you could punch, George."[50] By the eighth round, Foreman had punched himself out, with no energy to take the fight to his challenger. Ali sensed it was time to turn the tables, suddenly shifting from the pursued to the pursuer, pummeling Foreman with energetic combinations that eventually sent Foreman to the canvas. The referee stopped the fight with two seconds left in the eighth round, and Ali had defied all odds and regained the world heavyweight boxing title.

The Rumble in the Jungle is not really a story of Samson and Goliath, because Ali was still a very proficient fighter. The reason the match is notable in boxing circles, and is relevant to the topic at hand, is that it demonstrated an attribute of Muhammad Ali that arguably played a large factor in making him "The Greatest" boxer of his generation: he was a gifted tactician. This factor, in conjunction with his ample boxing skills, made him an unassailable opponent, even when it seemed that the odds were stacked against him.

In many ways, *Empowering the Great Energy Transition* is another manifestation of the Rumble in the Jungle. The opponent—a fossil fuel industry that has incomparable financial might and the power to alter market dynamics—is a daunting foe. If interests that favor renewable energy and smart-grid networks wish to ensure victory, they need to adopt more cohesive, targeted, and collaborative strategies. As our assessment of evolving market trends suggests, chances are they will prevail anyway; however, there is a lot to lose by allowing this battle to carry on into later rounds. The time for implementing the "rope a dope" for strategic effect has passed us by; all of us need to get off our stools beginning in the next available round and start swinging. We need to reset our goals and convictions so we can empower future generations and be part of something we can all be proud of.

Notes

1. The Great Energy Transition

1. International Energy Agency (IEA), *World Energy Outlook 2018*, https://www.iea .org/weo2018/.
2. "Bush: Kyoto Treaty Would Have Hurt US Economy," *NBC news online*, June 30, 2005, accessed April 18, 2014, http://www.nbcnews.com/id/8422343/ns/politics/t /bush-kyoto-treaty-would-have-hurt-economy/#.U1326FWSx8E.
3. "Harper's Letter Dismisses Kyoto as 'Socialist Scheme,'" *CBC News*, January 30, 2007, accessed April 28, 2014, http://www.cbc.ca/news/canada/harper-s-letter -dismisses-kyoto-as-socialist-scheme-1.693166.
4. P. Christoff, "Cold Climate in Copenhagen: China and the United States at COP15," *Environmental Politics* 19, no. 4 (2010): 637–656.
5. As reported by Lenore Taylor, "Renewable Energy: Tony Abbott Signals He Could Wind Back or Scrap Targets," *The Guardian*, December 17, 2013.
6. IEA, 2018.
7. British Petroleum, *BP Statistical Review of World Energy*, June 2017.
8. K. M. Campbell and J. Price, eds., *The Global Politics of Energy* (Washington, DC: The Aspen Institute, 2008).
9. Bethany McLean, *Saudi America: The Truth About Fracking and How It's Changing the World* (New York: Columbia Global Reports, 2018).
10. D. Yergin, "Energy Under Stress," in *The Global Politics of Energy*, ed. K. M. Campbell and J. Price (Washington, DC: The Aspen Institute, 2008), 27–43.
11. U.S. Energy Information Administration, *Total Energy*, accessed November 3, 2017, https://www.eia.gov/totalenergy/data/annual/showtext.php?t=ptb0709.

12. U.S. Energy Information Administration, *Coal Explained: Coal Prices and Outlook*, accessed December 14, 2018, https://www.eia.gov/energyexplained/index .php?page=coal_prices.

13. U.S. Energy Information Administration, *Electric Power Monthly with Data for November 2018*, Table ES.2.B: Summary Statistics: Receipts and Cost of Fossil Fuels for the Electric Power Industry by Sector, Btus, 2018 and 2017, accessed December 14, 2018, https://www.eia.gov/electricity/monthly/current_month/epm.pdf.

14. Data from the U.S. Energy Information Administration National Energy Modeling System 2018, operated by the Georgia Institute of Technology's Climate and Energy Policy Laboratory (www.cepl.gatech.edu).

15. Data from U.S. Energy Information Administration monthly energy price updates, accessed December 14, 2018, https://www.eia.gov/totalenergy.

16. U.S. Energy Information Administration, *Natural Gas*, accessed December 14, 2018, https://www.eia.gov/dnav/ng/hist/rngwhhdD.htm.

17. British Petroleum, 2017.

18. U.S. Energy Information Administration, *Natural Gas*, https://www.eia.gov/dnav /ng/ng_enr_shalegas_dcu_NUS_a.htm.

19. Robert Rapier, "How the Shale Boom Turned the World Upside Down," *Forbes*, April 21, 2017, accessed November 4, 2017, https://www.forbes.com/sites/rrapier/2017 /04/21/how-the-shale-boom-turned-the-world-upside-down/2/#1221cbe06c46.

20. International Energy Agency (IEA), *World Energy Outlook 2017* (Paris: IEA, 2017), 371.

21. EIA National Energy Modeling System 2018.

22. British Petroleum, 2017.

23. Yergin, 2008.

24. Rapier, 2017.

25. Rapier, 2017.

26. R. Johnson, "30 Million Sales in China by 2020? That Can't Happen, Right?" *Automotive News*, May 16, 2016, https://www.autonews.com/article/20160516/GLOBAL03 /305169935/30-million-sales-in-china-by-2020-that-can-t-happen-right.

27. Data taken from https://ycharts.com/indicators/crude_oil_spot_price, accessed November 4, 2017.

28. U.S. Energy Information Administration, *Natural Gas*, accessed December 14, 2018, https://www.eia.gov/dnav/ng/hist/n3010us3m.htm.

29. S. V. Valentine, "Emerging Symbiosis: Renewable Energy and Energy Security," *Renewable and Sustainable Energy Reviews* 15 (2011): 4572–4578.

30. J. Ho, "Taipower Deficit to Widen," *Taipei Times*, March 16, 2007, 11.

31. S. V. Valentine, "The Socio-Political Economy of Electricity Generation in China," *Renewable and Sustainable Energy Reviews* 32 (2014): 416–429.

32. S. V. Valentine, "Understanding the Variability of Wind Power Costs," *Renewable and Sustainable Energy Reviews* 15 (2011): 3632–3639.

33. J. Diamond, *Collapse: How Societies Choose to Fail or Survive* (London: Allen Lane, 2005).

34. D. Yergin, *The Prize: The Epic Quest for Oil, Money & Power* (New York: Free Press, 1993).

35. Yergin, 1993.

36. P. Roberts, *The End of Oil* (Boston: Houghton Mifflin, 2004).

37. Campbell and Price, 2008.

38. As reported by the editors of the *Washington Post*: "Europe Needs an Alternative to Russian Natural Gas," March 6, 2014.

39. V. Vivoda, "Japan's Energy Security Predicament Post-Fukushima," *Energy Policy* 46 (2012): 135–143.

40. M. I. Stein, "After Harvey, What Will Happen to Houston's Oil Industry?" *Wired*, September 21, 2017, accessed January 14, 2018, https://www.wired.com/story/what-will-happen-to-the-gulf-coast-if-the-oil-industry-retreats/.

41. James David Dickson and Holly Fournier, "730K Remain Without Power After Michigan Wind Storm," *Detroit News*, March 9, 2017, accessed November 4, 2018, http://www.detroitnews.com/story/news/local/michigan/2017/03/08/90k-without-power-and-winds-sweep-metro-detroit/98899718/.

42. "Central America Hit by Massive Power Outages," *Manila Times*, July 2, 2017, accessed November 4, 2017, http://www.manilatimes.net/central-america-hit-massive-power-outages/336160/.

43. T. L. Friedman, *Hot, Flat and Crowded* (New York: Farrar, Strauss and Giroux, 2008).

44. Friedman, 2008.

45. Friedman, 2008.

46. J. D. Sachs, *Common Wealth: Economics for a Crowded Planet* (New York: Penguin Press, 2008).

47. Council on Foreign Relations, *Terrorist Havens: Indonesia*, http://www.cfr.org/indonesia/terrorism-havens-indonesia/p9361.

48. R. Costanza et al., *An Introduction to Ecological Economics* (Boca Raton, FL: St. Lucie Press, 1997).

49. International Renewable Energy Agency (IRENA), *30 Years of Policies for Wind Energy: Lessons from 12 Wind Energy Markets* (Abu Dhabi, United Arab Emirates: IRENA, 2012).

50. For more on this study, see the State of Global Air website at https://www.stateofglobalair.org/.

51. F. Caiazzo et al., "Air Pollution and Early Deaths in the United States, Part I: Quantifying the Impact of Major Sectors in 2005," *Atmospheric Environment* 79 (2013): 198–208.

52. See the ExternE website: http://www.externe.info/externe_d7/?q=node/6, which details a comprehensive project designed to quantify external costs associated with energy generation.

53. https://flowcharts.llnl.gov/content/assets/images/charts/Energy/Energy_2017_United-States.png.

54. Brian Kahn, "Record Number of Billion-Dollar Disasters Globally in 2013," *Climate Central*, February 5, 2014, accessed April 29, 2014, http://www.climatecentral.org/news/globe-saw-a-record-number-of-billion-dollar-disasters-in-2013-17037.

55. Marshall Shepherd, "The Cost of Weather-Climate Disasters in 2017 Was $306 Billion—What Could You Buy with That?" *Forbes*, January 9, 2018, accessed December 14, 2018, https://www.forbes.com/sites/marshallshepherd/2018/01/09/cost-of-weather-climate-disasters-was-306-billion-in-2017-what-could-you-buy-with-that/#6006c57e71ed.

56. World Nuclear Association, *WNA Nuclear Century Outlook Data*, 2008, http://www.world-nuclear.org/our-association/publications/online-reports/nco/nuclear-century-outlook-data(1).aspx.

57. B. K. Sovacool and S. V. Valentine, *The National Politics of Nuclear Power: Economics, Security and Governance* (Milton Park, UK: Routledge, 2012).

58. "Three Years After: Experts Debate on Whether Japan Needs Nuclear Energy," *Asahi Shinbun*, March 6, 2014.

59. E. Johnston, "Balance of Power: Shift Toward Renewable Energy Appears to Be Picking Up Steam," *Japan Times*, October 14, 2017, accessed November 5, 2017, https://www.japantimes.co.jp/news/2017/10/14/business/balance-power-shift-toward-renewable-energy-appears-picking-steam/#.Wf6H-2iCzIU.

60. T. Osaki, "Junichiro Koizumi-Led Group Pitches Bill Calling for 'Immediate Halt' to Japan's Reliance on Nuclear Power," *Japan Times*, January 10, 2018.

61. However, it should be noted that under President Moon Jae-in, there appears to be an intention to move South Korea away from an energy policy dominated by nuclear power. The success of this policy will likely largely depend on the continued economic viability of renewable energy technologies and on Moon's ability to effectively oppose South Korea's nuclear power regime.

62. L. Richardson and M. Gurtov, "Protesting Policy and Practice in South Korea's Nuclear Energy Industry," *Asia-Pacific Journal* 15, no. 21 (2017), accessed November 5, 2017, http://apjjf.org/2017/21/Richardson.html.

63. World Nuclear Association, *Asia's Nuclear Energy Growth*, accessed January 14, 2018, http://www.world-nuclear.org/information-library/country-profiles/others/asias-nuclear-energy-growth.aspx.

64. P. A. Kharecha and J. E. Hansen, "Prevented Mortality and Greenhouse Gas Emissions from Historical and Projected Nuclear Power," *Environmental Science and Technology* 47 (2013): 4889–4895.

65. B. K. Sovacool, "A Critical Evaluation of Nuclear Power and Renewable Energy in Asia," *Journal of Contemporary Asia* 40, no. 3 (2010): 369–400.

66. J. W. Eerkens, *The Nuclear Imperative: A Critical Look at the Approaching Energy Crisis*, 2nd ed. (Heidelberg, Germany: Springer, 2010).

67. A. Galperin, P. Reichert, and A. Radkowsky, "Thorium Fuel for Light Water Reactors—Reducing Proliferation Potential of Nuclear Power Fuel Cycle," *Science and Global Security* 6, no. 3 (1997): 265–290; R. Hargraves and R. Moir, "Liquid Fluoride Thorium Reactors: An Old Idea in Nuclear Power Gets Reexamined," *American Scientist* 98, no. 4 (2010): 304–313.

68. B. K. Sovacool and C. Cooper, "Nuclear Nonsense: Why Nuclear Power Is No Answer to Climate Change and the World's Post-Kyoto Energy Challenges," *William*

and Mary Environmental Law and Policy Review 33, no. 1 (2008): 1–119; A. Gilbert et al., "Cost Overruns and Financial Risk in the Construction of Nuclear Power Reactors: A Critical Appraisal," *Energy Policy* 102 (2017): 644–649.

69. Marilyn A. Brown, "A Case for Completing Plant Vogtle," *Atlanta Journal Constitution*, September 1, 2017, http://www.myajc.com/news/opinion/opinion-case-for -completing-plant-vogtle/nGqXhaQ9OzKkWGofRh0tzM/.

70. Michael E. Webber, "Why the Withering Nuclear Power Industry Threatens US National Security," *The Conversation*, August 10, 2017, https://theconversation.com /why-the-withering-nuclear-power-industry-threatens-us-national-security-82351.

71. The report is available on the Greenpeace website at http://www.greenpeace.org /india/Global/india/docs/India-China-Air-Quality-Standards-Comparison.pdf, accessed April 30, 2014.

72. Source of data: *Lazard's Levelized Cost of Energy Analysis—Version 10.0*, accessed November 5, 2017, https://www.lazard.com/media/438038/levelized-cost-of-energy -v100.pdf.

73. Ran Fu et al., *U.S. Solar Photovoltaic System Cost Benchmark: Q1 2017*, National Renewable Energy Laboratory, prepared under Task No. SETP.10308.03.01.10, accessed November 5, 2017, https://www.nrel.gov/docs/fy17osti/68925.pdf.

74. Donald Chung et al., *U.S. Photovoltaic Prices and Cost Breakdowns: Q1 2015 Benchmarks for Residential, Commercial, and Utility-Scale Systems*, National Renewable Energy Laboratory, Technical Report NREL/TP-6A20-64746, accessed November 5, 2017, https://www.nrel.gov/docs/fy15osti/64746.pdf.

75. R. Andreas Kraemer, "The Death of Fossil Fuels," Twitter, December 10, 2017, tweeted from International Renewable Energy Agency.https://twitter.com/rakraemer.

76. Steve Hargreaves, "Energy Subsidies Total $24 Billion, Most to Renewables," *CNN Money*, March 7, 2012, accessed April 30, 2014, http://money.cnn.com/2012/03/07 /news/economy/energy-subsidies/.

77. Renewable Energy Policy Network for the 21st Century (REN21), *Renewables 2017 Global Status Report*, http://www.ren21.net/gsr-2017/.

78. Renewable Energy Policy Network for the 21st Century (REN21), *Renewables 2018 Global Status* Report, http://www.ren21.net/gsr-2018/.

79. REN21, 2018.

80. Federal Ministry for the Environment, Nature Conservation and Nuclear Safety, "Environmental Technology Atlas for Germany," *GreenTech Made in Germany*, https://www.greentech-made-in-germany.de/en/.

81. The contents of this speech were available on the Whitehouse website, http://www.white house.gov/the_press_office/Clean-Energy-Economy-Fact-Sheet, accessed April 30, 2014. However, the link is now dead.

82. Remarks by President Obama at the National Clean Energy Summit, Las Vegas, Nevada, August 25, 2015. Speech available online at https://www.whitehouse.gov/the -press-office/2015/08/25/remarks-president-national-clean-energy-summit.

83. M. Richtell, "Start-Up Fervor Shifts to Energy in Silicon Valley," *New York Times*, March 14, 2007, https://www.nytimes.com/2007/03/14/technology/14valley.html.

84. D. C. Mueller, "First-Mover Advantages and Path Dependence," *International Journal of Industrial Organization* 15 (1997): 827–850.

85. P. Doyle, *Marketing Management and Strategy*, 2nd ed. (London: Prentice Hall Europe, 1998).

86. Federal Ministry for the Environment, Nature Conservation and Nuclear Safety, 2011.

87. REN21, 2018.

2. Sneak Preview: Challenges Underpinning the Energy Transition

1. Travis Bradford, "A Brief History of Energy," in *Solar Revolution: The Economic Transformation of the Global Energy Industry* (Boston: MIT Press, 2006), 23–43.

2. N. Stern, *The Stern Review: Report on the Economics of Climate Change* (London: Cabinet Office—HM Treasury, 2006).

3. IPCC, 2014: *Climate Change 2014: Mitigation of Climate Change. Contribution of Working Group III to the Fifth Assessment Report of the Intergovernmental Panel on Climate Change*, ed. O. Edenhofer et al. (Cambridge: Cambridge University Press, 2014).

4. IPCC, 2014, 8.

5. See an analysis at climatecentral.org: http://www.climatecentral.org/news/major-greenhouse-gas-reductions-needed-to-curtail-climate-change-ipcc-17300.

6. J. Hansen et al., "Target Atmospheric CO_2: Where Should Humanity Aim?" *Open Atmospheric Science Journal* 2 (2008): 217–231.

7. M. B. McElroy, *The Atmospheric Environment: Effects of Human Activity* (Princeton, NJ: Princeton University Press, 2002).

8. Stern, 2006.

9. IPCC, 2014.

10. D. Yergin, *The Prize: The Epic Quest for Oil, Money & Power* (New York: Free Press, 1993).

11. Yergin, 1993.

12. T. Kuhn, *The Structure of Scientific Revolutions*, 3rd ed. (Chicago: University of Chicago Press, 1996).

13. ExxonMobil's annual reports are available at https://corporate.exxonmobil.com/en/investors/annual-report.

14. This article can be accessed via https://www.investopedia.com/ask/answers/030915/what-percentage-global-economy-comprised-oil-gas-drilling-sector.asp.

15. Bradford, 2006.

16. Renewable Energy Policy Network for the 21st Century (REN21), *Renewables 2018 Global Status Report*, http://www.ren21.net/gsr-2018/.

17. International Energy Agency (IEA) website: https://www.iea.org/publications/wei2017/.

18. B. Sovacool, M. A. Brown, and S. Valentine, *Fact and Fiction in Global Energy Policy: Fifteen Contentious Questions* (Baltimore, MD: Johns Hopkins University Press, 2016), 337.

19. Sovacool, Brown, and Valentine, 2016.
20. D. Coady et al., "How Large Are Global Fossil Fuel Subsidies?" Supplement C, *World Development* 91(2017): 11–27.
21. N. Oreskes and E. M. Conway, *Merchants of Doubt* (London: Bloomsbury Press, 2010).
22. J. Hansen, "Global Warming Twenty Years Later: Tipping Points Near" (paper presented at the National Press Club, Washington, DC, June 23, 2008).
23. Robert Brulle, "Institutionalizing Delay: Foundation Funding and the Creation of U.S. Climate Change Counter-Movement Organizations," *Climatic Change* 122, no. 4 (2014): 681–694.
24. Robert J. Brulle, Liesel Hall Turner, Jason Carmichael, and J. Craig Jenkins, "Measuring Social Movement Organization Populations: A Comprehensive Census of U.S. Environmental Movement Organizations," *Mobilization: An International Quarterly Review* 12, no. 3 (2007): 195–211; Justin Farrell, "Corporate Funding and Ideological Polarization About Climate Change," *Proceedings of the National Academy of Sciences* 113, no. 1 (2016): 92–97; Justin Farrell, "Network Structure and Influence of the Climate Change Counter-Movement," *Nature Climate Change* 6, no. 4 (2016): 370–374; Peter J. Jacques, Riley E. Dunlap, and Mark Freeman, "The Organisation of Denial: Conservative Think Tanks and Environmental Scepticism," *Environmental Politics* 17, no. 3 (2008): 349–385; Aaron M. McCright and Riley E. Dunlap, "Defeating Kyoto: The Conservative Movement's Impact on U.S. Climate Change Policy," *Social Problems* 50, no. 3 (2003): 348–373.
25. National Academy of Sciences, *Climate Intervention: Carbon Dioxide Removal and Reliable Sequestration* (Washington, DC: National Academies Press, 2015), http://nap.edu/18805.
26. Thomas Vangkilde-Pedersen et al., "Assessing European Capacity for Geological Storage of Carbon Dioxide: The EU GeoCapacity Project," *Energy Procedia* 1, no. 1 (2009): S21.
27. B. Sovacool, M. A. Brown, and S. Valentine, *Fact and Fiction in Global Energy Policy: Fifteen Contentious Questions* (Baltimore, MD: Johns Hopkins University Press, 2016).
28. B. Elliston, I. MacGill, and M. Diesendorf, "Least Cost 100 Percent Renewable Electricity Scenarios in the Australian National Electricity Market," *Energy Policy* 59 (2013): 270–282.
29. Wu, Jing, Botterud Audun, Mills Andrew, Zhou Zhi, Hodge Bri-Mathias, and Mike Heaney. "Integrating solar PV (photovoltaics) in utility system operations: Analytical framework and Arizona case study." *Energy*, 85, (2015): 1-9.
30. Council of Economic Advisors (CEA), "Incorporating Renewables into the Electric Grid: Expanding Opportunities for Smart Markets and Energy Storage" (June 2016).
31. Lu, S. et. al. 2011. Large Scale PV Integration Study. Publication. Pacific Northwest National Laboratory. 20677-PNNL, accessed August 2017.
32. Evaluation taken from Vivid Economics. 2016. "Money to Burn? The U.K. needs to dump biomass and replace its coal plants with truly clean energy." Issue Brief,

Natural Resources Defense Council and Vivid Economics. 2017. "Money to Burn II." Issue Brief, Natural Resources Defense Council.

33. Brown, M.A., A. Favero, V.M. Thomas, and A. Banboukian. (2019) "The Economic and Environmental Performance of Biomass Power as an Intermediate Resource for Power Production," Utilities Policy, 58: 52–62.

34. C. T. Clack et al., "Evaluation of a Proposal for Reliable Low-Cost Grid Power with 100 Percent Wind, Water, and Solar," *Proceedings of the National Academy of Sciences* 114, no. 26 (2017), https://doi.org/10.1073/pnas.1610381114; U.S. Department of Energy, *Staff Report to the Secretary on Electricity Markets and Reliability*, 2017, https://www.eenews.net/assets/2017/08/24/document_gw_06.pdf.

35. Source of data: *Lazard's Levelized Cost of Energy Analysis—Version 10.0*, accessed November 5, 2017, https://www.lazard.com/media/438038/levelized-cost-of-energy-v100.pdf.

36. U.S. Department of Energy, 2017, 61–101. This report has an interesting discussion of the many definitions of "premature retirement" of a power plant.

37. S. V. Valentine, *Wind Power Politics and Policy* (Oxford: Oxford University Press, 2014).

38. S. V. Valentine, "Canada's Constitutional Separation of (Wind) Power," *Energy Policy* 38, no. 4 (2010): 1918–1930.

39. S. V. Valentine, "The Socio-Political Economy of Electricity Generation in China," *Renewable and Sustainable Energy Reviews* 32 (2014): 416–429.

40. Valentine, "The Socio-Political Economy," 2014.

41. C. Christensen, *The Innovator's Dilemma: When New Technologies Cause Great Firms to Fail* (Boston: Harvard Business School Press, 2013).

42. Christensen, 2013.

43. W. J. Nuttall, *Nuclear Renaissance: Technologies and Policies for the Future of Nuclear Power* (Boca Raton, FL: CRC Press, 2004); A. Stulberg and M. Fuhrmann, *The Nuclear Renaissance and International Security* (Stanford, CA: Stanford University Press, 2013).

44. International Energy Agency (IEA), *World Energy Outlook 2016* (Paris: IEA, 2016).

45. See the World Nuclear Association website: http://www.world-nuclear.org/information-library/country-profiles/others/asias-nuclear-energy-growth.aspx.

46. B. K. Sovacool, *Contesting the Future of Nuclear Power* (Singapore: World Scientific Publishing, 2011).

47. B. K. Sovacool and S. V. Valentine, *The National Politics of Nuclear Power: Economics, Security and Governance* (Milton Park, UK: Routledge, 2012).

48. See the World Nuclear Association website: http://www.world-nuclear.org/information-library/country-profiles/others/asias-nuclear-energy-growth.aspx.

49. Sovacool and Valentine, 2012.

50. S. Valentine, "Braking Wind in Australia: A Critical Evaluation of the Renewable Energy Target," *Energy Policy* 38, no. 7 (2010): 3668–3675.

51. Valentine, "The Socio-Political Economy," 2014.

52. Valentine, *Wind Power Politics*, 2014.

53. E. Lantz, R. Wiser, and M. Hand, *IEA Wind Task 26: The Past and Future Cost of Wind Energy*, Work Package 2: National Renewable Energy Laboratory (NREL), 2012, Golden, CO.

54. For more information from the laboratory that is developing this technology, go to https://www.riam.kyushu-u.ac.jp/windeng/img/aboutus_detail_image/Wind_Lens_20140601.pdf.

55. International Renewable Energy Agency (IRENA), *Renewable Power Generation Costs in 2014* (Bonn, Germany: IRENA, 2015).

56. R. R. Lunt and V. Bulovic, "Transparent, Near-Infrared Organic Photovoltaic Solar Cells for Window and Energy-Scavenging Applications," *Applied Physics Letters* 98, no. 11 (2011): 113305.

57. See the website of Ubiquitous Energy: http://ubiquitous.energy/technology/.

58. More on this can be found at http://inhabitat.com/ucla-doubles-efficiency-of-transparent-solar-power-generating-film/.

59. For more on this story, see www.tesla.com/blog/tesla-and-solarcity.

60. R. Rosenblatt, *Consuming Desires: Consumption, Culture, and the Pursuit of Happiness* (Washington, DC: Island Press, 2006).

61. J. G. Speth, *The Bridge at the End of the World: Capitalism, the Environment, and Crossing from Crisis to Sustainability* (New Haven, CT: Yale University Press, 2008).

62. N. Carter, *The Politics of the Environment: Ideas, Activism and Policy* (Cambridge: Cambridge University Press, 2004).

63. Paul N. Edwards, "Infrastructure and Modernity: Force, Time, and Social Organization in the History of Sociotechnical Systems," in *Modernity and Technology*, ed. Thomas J. Misa, Philip Brey, and Andrew Feenberg (Cambridge, MA: MIT Press, 2003), 185–186.

64. Sovacool, Brown, and Valentine, 2016.

65. E. M. Rogers, *Diffusion of Innovations*, 4th ed. (New York: Free Press, 1995).

66. Valentine, "The Socio-Political Economy," 2014.

3. The Uncertainties of Climate Change

1. "Whatever Happened to Global Warming? How Freezing Temperatures Are Starting to Shatter Climate Change Theory," *Daily Mail*, October 14, 2009, accessed October 20, 2017, https://www.dailymail.co.uk/sciencetech/article-1220052/Austria-sees-earliest-snow-history-America-sees-lowest-temperatures-50-years-So-did-global-warming-go.html.

2. For a highly readable explanation of the connection between cold spells and climate change, see Bryan Walsh, "Climate Change Might Just Be Driving the Historic Cold Snap," *Time*, January 6, 2014, http://science.time.com/2014/01/06/climate-change-driving-cold-weather/.

3. IPCC, 2014: *Climate Change 2014: Synthesis Report. Contribution of Working Groups I, II and III to the Fifth Assessment Report of the Intergovernmental Panel on Climate*

Change, ed. R. K. Pachauri and L. A. Meyer (Geneva, Switzerland: IPCC, 2014), 151 pp.

4. David Whitehouse, "Sunspots Reaching 1,000-Year High," *BBC News*, July 6, 2004, accessed October 20, 2017, http://news.bbc.co.uk/2/hi/science/nature/3869753.stm.

5. David Gamble, "How Do We Know That Humans Are Responsible for Climate Change?" *Skeptical Science*, August 25, 2018, accessed May 3, 2019, https://skeptical-science.com/argument.php.

6. Ian Plimer, "Legislative Time Bomb," *ABC News Australia*, September 28, 2010, accessed October 20, 2017, http://www.abc.net.au/news/2009-08-13/29320.

7. G. A. Meehl et al., "Combinations of Natural and Anthropogenic Forcings in Twentieth -Century Climate," *Journal of Climate* 17 (2004): 3721–3727.

8. Meehl et al., 2004.

9. Robert Mendelsohn, "Can We Survive Climate Change? The Critical Role of Adaptation," November 29, 2018. https://cepl.gatech.edu/sites/default/files/attachments/robert-mendelsohn-gatech-seminar-2018-11-29.pdf.

10. M. Burke, S. M. Hsiang, and E. Miguel, "Global Non-Linear Effect of Temperature on Economic Production," *Nature* 527, no. 7577 (2015): 235–239.

11. Paul Waldie, "A Reality Check on the Northwest Passage Boom," *The Globe and Mail*, January 7, 2014.

12. More on this study can be found at http://nrt-trn.ca/climate/climate-prosperity/the-economic-impacts-of-climate-change-for-canada/paying-the-price.

13. B. J. Bentz et al., "Climate Change and Bark Beetles of the Western United States and Canada: Direct and Indirect Effects," *BioScience* 60, no. 8 (2010): 602–613.

14. For more on this, see "500 Scientists Refute Global Warming Dangers," accessed October 23, 2017, http://www.wnd.com/2007/09/43489/.

15. R. Warren et al., "Quantifying the Benefit of Early Climate Change Mitigation in Avoiding Biodiversity Loss," *Nature Climate Change* 3 (2013): 678–682.

16. R. Black, D. Kniveton, and K. Schmidt-Verkerk, "Migration and Climate Change: Toward an Integrated Assessment of Sensitivity," in *Disentangling Migration and Climate Change*, ed. T. Faist and J. Schade (Heidelberg, Germany: Springer Netherlands, 2013), 29–53.

17. M. Brzoska and C. Fröhlich, "Climate Change, Migration and Violent Conflict: Vulnerabilities, Pathways and Adaptation Strategies," *Migration and Development* 5, no. 2 (2016): 190–210.

18. Franklin Hadley Cocks, "Global Warming vs. the Next Ice Age," *MIT Technology Review*, 2009, accessed October 23, 2017, https://www.technologyreview.com/s/416786/global-warming-vs-the-next-ice-age/.

19. An excellent site that refutes many of the more prominent climate change denials is www.skepticalscience.com.

20. Edward Wong, "Trump Has Called Climate Change a Chinese Hoax. Beijing Says It Is Anything But," *New York Times*, November 18, 2016, accessed October 23, 2017, https://www.nytimes.com/2016/11/19/world/asia/china-trump-climate-change.html.

21. "EPA Head Scott Pruitt Denies That Carbon Dioxide Causes Global Warming," *The Guardian*, March 9, 2017, accessed October 20, 2017, https://www.theguardian .com/environment/2017/mar/09/epa-scott-pruitt-carbon-dioxide-global-warming -climate-change.

22. JoAnna Wendel, "Unseasonable Weather Entrenches Climate Opinions," *Eos*, May 24, 2017, accessed October 20, 2017, https://eos.org/articles/unseasonable-weather -entrenches-climate-opinions.

23. Intergovernmental Panel on Climate Change website, accessed May 3, 2019, https:// www.ipcc.ch/about/structure/.

24. J. Hinkel et al., "Coastal Flood Damage and Adaptation Costs Under 21st Century Sea-Level Rise," *Proceedings of the National Academy of Sciences (PNAS)* 111, no. 9 (2014): 3292–3297.

25. C. Tebaldi, B. H. Strauss, and C.E. Zervas, "Modelling Sea Level Rise Impacts on Storm Surges Along US Coasts," *Environmental Research Letters* 7, no. 1: 014032.

26. Intergovernmental Panel on Climate Change website, accessed May 3, 2019, https:// www.ipcc.ch/about/structure/.

27. IPCC, 2014.

28. Alan Buis, "Is a Sleeping Climate Giant Stirring in the Arctic?" NASA, accessed October 25, 2017, https://www.nasa.gov/topics/earth/features/earth20130610.html.

29. M. D. Cooper et al., "Limited Contribution of Permafrost Carbon to Methane Release from Thawing Peatlands," *Nature Climate Change* 7, no. 7 (2017): 507.

30. R. S. Williams, Jr., J. G. Ferrigno, and K. M. Foley, *Coastal-Change and Glaciological Maps of Antarctica*, U.S. Geological Survey Fact Sheet 2005–3055, https://pubs.usgs .gov/fs/2005/3055. See also http://www.antarcticglaciers.org/glaciers-and-climate /estimating-glacier-contribution-to-sea-level-rise/.

31. IPCC, 2013: *Climate Change 2013: The Physical Science Basis. Contribution of Working Group I to the Fifth Assessment Report of the Intergovernmental Panel on Climate Change*, ed. T. F. Stocker et al. (Cambridge: Cambridge University Press, 2013), 1535 pp.

32. W. S. Broecker, "Was the Younger Dryas Triggered by a Flood?" *Science* 312, no. 5777 (2006): 1146–1148.

33. IPCC, 2014.

34. N. H. Stern, *The Economics of Climate Change: The Stern Review* (Cambridge: Cambridge University Press, 2007), i.

35. M. Springmann et al., "Global and Regional Health Effects of Future Food Production Under Climate Change: A Modelling Study," *The Lancet* 387, no. 10031 (2016): 1937–1946.

36. World Health Organization, *Climate Change and Health*, accessed October 25, 2017, http://www.who.int/mediacentre/factsheets/fs266/en/.

37. Travis Bradford, "A Brief History of Energy," in *Solar Revolution: The Economic Transformation of the Global Energy Industry* (Boston: MIT Press, 2006), 23–43.

38. Quoted in B. K. Sovacool, L. D. Noel, and R. J. Orsato, "Stretching, Embeddedness, and Scripts in a Sociotechnical Transition: Explaining the Failure of Electric

Mobility at Better Place (2007–2013)," *Technological Forecasting and Social Change* 123 (October 2017): 24–34.

39. https://www.marketwatch.com/investing/stock/xom/financials, accessed November 15, 2017.

40. GDP estimates from the World Bank, accessed November 25, 2017, http://databank .worldbank.org/data/download/GDP.pdf.

41. https://www.marketwatch.com/investing/stock/rds.a/financials, accessed November 15, 2017.

42. BP website, accessed November 15, 2017, https://www.bp.com/en/global/corporate /who-we-are.html.

43. Toyota website, accessed November 15, 2017, http://corporatenews.pressroom.toyota .com/releases/toyota-april-march-2016-financial-results.htm.

44. http://www.nasdaq.com/symbol/xom/financials?query=balance-sheet, accessed November 17, 2017.

45. Data from www.marketwatch.com and www.nasdaq.com, accessed November 17, 2017.

46. Data accessed November 17, 2017, at http://thehill.com/business-a-lobbying/business -a-lobbying/318177-lobbyings-top-50-whos-spending-big.

47. Julia Pyper, "Shell Plans to Boost Clean Energy Spending to $1 Billion per Year," *greentech-media.com*, July 11, 2017, accessed November 17, 2017, https://www.greentechmedia .com/articles/read/shell-boost-clean-energy-spending-1-billion-2020#gs.0sx92t4.

48. Blagojce Krivevski, "Global Plug-in Vehicle Sales for 2016," *Electric Cars Report*, accessed November 19, 2018, https://electriccarsreport.com/2017/02/global-plug-vehicle-sales -2016/.

49. https://www.volkswagenag.com/en/news/2018/03/VolkswagenGroup_expand _production.html.

50. *Global Warming Twenty Years Later: Tipping Points Near: Briefing to the House Select Comm. on Energy Independence and Global Warming* (2008) (statement of J. Hansen, Director of the NASA Goddard Institute for Space Studies).

51. K. Mulvey and S. Shulman, *The Climate Deception Dossiers: Internal Fossil Fuel Industry Memos Reveal Decades of Corporate Disinformation*, Union of Concerned Scientists, 2015, https://www.ucsusa.org/sites/default/files/attach/2015/07/The-Climate -Deception-Dossiers.pdf.

52. Kevin Kalhoefer, "STUDY: CNN Viewers See Far More Fossil Fuel Advertising than Climate Change Reporting," *MediaMatters*, April 25, 2016, accessed November 17, 2017, https://www.mediamatters.org/research/2016/04/25/study-cnn-viewers-see-far -more-fossil-fuel-advertising-climate-change-reporting/209985.

53. Mark Baer, "This Presidential Election Reflects the Dumbing Down of America More than Anything Else," *Huffington Post*, May 14, 2016, accessed November 17, 2017, https://www.huffingtonpost.com/entry/this-presidential-election-reflects-the -dumbing-down_us_57ebc343e4b07f20daa0ff89.

54. Amitava Banerjee, "A Call to Arms (or Pens) Against Anti-Intellectualism," *Huffington Post*, June 30, 2016, accessed November 17, 2017, http://www.huffingtonpost .co.uk/amitava-banerjee/eu-referendum-anti-intellectualism_b_10749140.html.

55. "EU Greenhouse Gas Emissions at Lowest Level Since 1990," *European Environment Agency*, June 21, 2016, accessed November 17, 2017, https://www.eea.europa.eu /highlights/eu-greenhouse-gas-emissions-at.

56. T. Ding, Y. Ning, and Y. Zhang, "Estimation of Greenhouse Gas Emissions in China 1990–2013," *Greenhouse Gases Science and Technology* 7, no. 6 (2017): 1097–1115, https://doi.org/10.1002/ghg.1718.

57. http://www.worldometers.info/world-population/.

58. http://www.un.org/sustainabledevelopment/blog/2015/07/un-projects-world -population-to-reach-8-5-billion-by-2030-driven-by-growth-in-developing-countries/.

59. D. Yergin, "Energy Under Stress," in *The Global Politics of Energy*, ed. K. M. Campbell and J. Price (Washington, DC: The Aspen Institute, 2008), 27–43.

60. V. Bellassen and S. Luyssaert, "Carbon Sequestration: Managing Forests in Uncertain Times," *Nature* 506, no. 7487 (2014): 153–155.

61. For a summary of what was achieved, see http://unfccc.int/meetings/copenhagen _dec_2009/meeting/6295.php, accessed November 20, 2017.

62. More on the Paris Agreement can be found at https://unfccc.int/process/the-paris -agreement/status-of-ratification, accessed July 17, 2018.

63. "France Set to Ban Sale of Petrol and Diesel Vehicles by 2040," *BBC News*, July 6, 2017, accessed November 20, 2017, http://www.bbc.com/news/world-europe-40518293.

64. More on EU efforts can be found at https://ec.europa.eu/clima/policies/strategies /2030_en, accessed November 20, 2017.

65. Oliver Milman, "James Hansen, Father of Climate Change Awareness, Calls Paris Talks 'a Fraud,'" *The Guardian*, December 12, 2015, accessed November 20, 2017, https://www.theguardian.com/environment/2015/dec/12/james-hansen-climate -change-paris-talks-fraud.

66. Laurel Hamers, "Global Carbon Dioxide Emissions Will Hit a Record High in 2018," *ScienceNews*, December 6, 2018, https://www.sciencenews.org/article/global -carbon-dioxide-emissions-will-hit-record-high-2018.

67. "AccuWeather Predicts Economic Cost of Harvey, Irma to Be $290 Billion," *AccuWeather*, September 11, 2017, accessed November 20, 2017, https://www .accuweather.com/en/weather-news/accuweather-predicts-economic-cost-of -harvey-irma-to-be-290-billion/70002686.

68. More on this story at https://www.ceres.org/annual-report/2016/progress/carbon -asset-risk-makes-major-headway, accessed May 5, 2019.

69. More on this story at https://internationaldirector.com/the-c-suite/oil-companies -pressure-measure-impact-policies-climate-change-businesses/, accessed May 3, 2019.

70. T. Kuhn, *The Structure of Scientific Revolutions*, 3rd ed. (Chicago: University of Chicago Press, 1996).

71. World Digital Library at https://www.wdl.org/en/item/3164/, accessed November 20, 2017.

72. See http://www.history.com/this-day-in-history/galileo-is-convicted-of-heresy at history.com, accessed November 20, 2017.

73. An interesting discussion of the affair is found at https://thonyc.wordpress. com/2017/09/28/galileo-the-church-and-that-ban/, accessed November 20, 2017.

74. "Pope Francis Warns 'History Will Judge' Climate Change Deniers," *Voc.com*, September 11, 2017, accessed October 20, 2017, https://www.vox.com/identities /2017/9/11/16290546/pope-francis-climate-change-deniers-daca.

4. Managing Uncertainties While Promoting Technological Evolution

1. https://boingboing.net/2008/01/03/roy-amara-forecaster.html.

2. C. Martin, F. Starace, and J. Pascal Tricoire, *The Future of Electricity: New Technologies Transforming the Grid Edge* (New York: World Economic Forum, 2017), http:// www3.weforum.org/docs/WEF_Future_of_Electricity_2017.pdf.

3. DNV-GL, *Energy Transition Outlook 2017*, https://eto.dnvgl.com/2017/main-report.

4. The *Internet of Things* refers to the emergent benefits from connected technological devices that allow data to be exchanged, resulting in efficiency improvements, economic benefits, and overall utilitarian benefits to society.

5. DNV-GL, 2017.

6. Additive manufacturing is computer-aided fabrication involving techniques such as the precision melting of powder materials using electron beams, the use of ultrasonic systems, and material addition with laser metal deposition. Additive manufacturing is all-electric and creates products with little material waste. See Marilyn A. Brown and Kim Gyungwon, "Energy and Manufacturing: Technology and Policy Transformations and Challenges," in *Handbook of Manufacturing Industries in the World Economy*, ed. John R. Bryson, Jennifer Clark, and Vida Vanchan (Cheltenham, UK: Edward Elgar, 2015), 121–146, http://www.elgaronline.com/view /9781781003923.00019.xml.

7. International Energy Agency, *Energy Access Outlook 2017*, October 19, 2017, https:// www.iea.org/access2017/.

8. R. Lowe, "Technical Options and Strategies for Decarbonizing UK Housing," *Building Research & Information* 35, no. 4 (2007): 412–425, doi:10.1080/09613210701238268; M. Sugiyama, "Climate Change Mitigation and Electrification," *Energy Policy* 44 (2012): 464–468, doi:10.1016/j.enpol.2012.01.028; Oswaldo Lucon and Diana Urge-Vorsatz, eds., "Buildings," in IPPC, 2014: *Climate Change 2014: Mitigation of Climate Change. Contribution of Working Group III to the Fifth Assessment Report of the Intergovernmental Panel on Climate Change*, ed. O. Edenhofer et al. (Cambridge: Cambridge University Press, 2014).

9. J. Cockroft and N. Kelly, "A Comparative Assessment of Future Heat and Power Sources for the UK Domestic Sector," *Energy Conversion and Management* 47, no. 15 (2006): 2349–2360, doi:10.1016/j.enconman.2005.11.021.

10. https://energy.gov/eere/articles/cold-climate-heat-pumps-help-consumers-stay -comfortable-and-save-money.

11. http://www.neep.org/initiatives/high-efficiency-products/emerging-technologies /ashp/cold-climate-air-source-heat-pump.

12. IPCC, 2014.

13. Renewable Energy Policy Network for the 21st Century (REN21), *Renewables 2018 Global Status Report*, http://www.ren21.net/gsr-2018/.

14. IPCC, 2014.

15. M. A. Brown et al., "Modeling Climate-Driven Changes in U.S. Buildings Energy Demand," *Climatic Change* 134, nos. 1–2 (2016): 29–44, doi:10.1007/s10584-015-1527-7.

16. Maximilian Auffhammer, Patrick Baylis, and Catherine H. Hausman, "Climate Change Is Projected to Have Severe Impacts on the Frequency and Intensity of Peak Electricity Demand Across the United States," *PNAS* 114, no. 8 (2017): 1886–1891, doi:10.1073/pnas.1613193114.

17. S. W. Hadley et al., "Responses of Energy Use to Climate Change: A Climate Modeling Study," *Geophysical Research Letters* 33, no. 17 (2006): L17703, doi:10.1029/2006GL026652.

18. Marilyn A. Brown et al., "Climate Change and Energy Demand in Buildings," *American Council for an Energy-Efficient Economy (ACEEE) Summer Study on Energy Efficiency in Buildings*, Pacific Grove, CA, 2014, accessed May 4, 2019, https://aceee.org/files/proceedings/2014/data/papers/3-736.pdf; E. T. Mansur, R. Mendelsohn, and W. Morrison, "Climate Change Adaptation: A Study of Fuel Choice and Consumption in the US Energy Sector," *Journal of Environmental Economics and Management* 55, no. 2 (2008): 175–193, http://dx.doi.org/10.1016/j.jeem.2007.10.001; M. J. Scott and Y. J. Huang, "Effects of Climate Change on Energy Use in the United States," in *Effects of Climate Change on Energy Production and Use in the United States*, report by the U.S. Climate Change Science Program and the subcommittee on Global Change Research, Washington, DC, 2007, 8–44.

19. U.S. Environmental Protection Agency (EPA), *The Potential Effects of Global Climate Change on the United States* (Washington, DC: EPA, 1989).

20. J. A. Sathaye et al., "Estimating Impacts of Warming Temperatures on California's Electricity System," *Global Environmental Change* 23, no. 2 (2013): 499–511, doi:10.1016/j.gloenvcha.2012.12.005.

21. REN21, 2018.

22. International Energy Agency website: http://www.iea.org/energyaccess/, accessed November 22, 2017.

23. H. Farhangi, "The Path of the Smart Grid," *IEEE Power and Energy Magazine* 8, no. 1 (2010): 18–28.

24. Memorandum by the Electricity Advisory Committee, "Distributed Energy Resource Valuation & Integration," 2017, U.S. Department of Energy, Washington, DC.

25. U.S. Energy Information Administration, *EIA Form 826—Net Metering 2016*. Data available through the website: https://www.eia.gov/electricity/data/eia861m/

26. U.S. Department of Energy (DOE), *Combined Heat and Power (CHP) Technical Potential in the United States*, 2016.

27. U.S. Energy Information Administration, *EIA Form 826—Small Scale Solar 2014 and February 2017*. Data available through the website: https://www.eia.gov/electricity/data/eia861m/

28. Federal Energy Regulatory Commission, *Assessment of Demand Response and Advanced Metering*, Staff Report, December 2016, https://www.ferc.gov/legal/staff-reports/2016/DR-AM-Report2016.pdf.

29. Johana Bhuiyan, "There Have Now Been over 540,000 Electric Vehicles Sold in the U.S.," *Recode*, December 21, 2016, accessed November 6, 2017, https://www.recode.net/2016/12/21/14041112/electric-vehicles-report-2016.

30. Examples include Consolidated Edison's Brooklyn Queens Demand Management program and Southern California Edison's Preferred Resources Pilot.

31. The term *co-benefits* refers to the positive side effects, secondary benefits, collateral benefits, or associated benefits from a particular green policy or clean energy system. Akiko Miyatsuka and Eric Zusman, *What Are Co-benefits?* Asian Co-benefits Partnership (ACP), October 2010, http://pub.iges.or.jp/modules/envirolib/upload/3378/attach/acp_factsheet_1_what_co-benefits.pdf.

32. PwC Global Power & Utilities, *Electricity Beyond the Grid: Accelerating Access to Sustainable Power for All*, May 2016, 6, https://www.pwc.com/gx/en/energy-utilities-mining/pdf/electricity-beyond-grid.pdf.

33. International Finance Corporation (IFC), *Off-Grid Solar Market Trends Report 2016* (Washington, DC: Bloomberg New Energy Finance and Lighting Global, 2016), 6, http://www.energynet.co.uk/webfm_send/1690.

34. Andrew Meyer, "Why a Distributed Energy Grid Is a Better Energy Grid," *Swell*, accessed February 14, 2017, https://www.swellenergy.com/blog/2016/05/20/why-a-distributed-energy-grid-is-a-better-energy-grid.

35. Kartikeya Singh, "Business Innovation and Diffusion of Off-Grid Solar Technologies in India," *Energy for Sustainable Development* 30 (2016): 1–13.

36. Siddharth Suryanarayanan and Elias Kyriakides, *Microgrids: An Emerging Technology to Enhance Power System Reliability*, IEEE Smart Grid Resource Center, March 12, 2012, http://resourcecenter.smartgrid.ieee.org/sg/product/publications/SGNL0057.

37. World Health Organization, *Household Air Pollution and Health*, May 8, 2018, http://www.who.int/mediacentre/factsheets/fs292/en/; N. D. Rao, A. Agarwal, and D. Wood, *Impacts of Small-Scale Electricity Systems: A Study of Rural Communities in India and Nepal* (Washington, DC: World Resources Institute, 2016), 38–39, http://pure.iiasa.ac.at/12913/.

38. IPCC, 2011: *Renewable Energy Sources and Climate Change Mitigation: Summary for Policymakers and Technical Summary, Special Report of the Intergovernmental Panel on Climate Change*, ed. O. Edenhofer et al. (Cambridge: Cambridge University Press, 2011), 65, https://www.ipcc.ch/pdf/special-reports/srren/SRREN_FD_SPM_final.pdf.

39. Clean Cooking Alliance, *Fighting Deforestation with Cleaner Cookstoves and Fuels*, November 2, 2016, accessed February 14, 2017, http://cleancookstoves.org/about/news/11-02-2016-fighting-deforestation-with-cleaner-cookstoves-and-fuels.html.

40. Yongping Zhai, "4 Ways to Empower Asian Women Through Energy Access," *Asian Development Blog*, January 12, 2016, accessed February 14, 2017, https://blogs.adb.org/blog/4-ways-empower-asian-women-through-energy-access; Katherine Lucey, "Women's Energy Entrepreneurship: Empowering Women and Scaling Access to

Energy," *Energia*, March 29, 2016, accessed May 4, 2019, https://www.energia.org /cws-60-recap-womens-energy-entrepreneurship-empowering-women-and-scaling -access-to-energy/.

41. Msolo Onditi, Irene Garcia, and Anna Leidreiter, *100 Percent Renewable Energy and Poverty Reduction in Tanzania*, April 2016, 7, https://www.worldfuturecouncil.org /wp-content/uploads/2016/04/WFC_CAN_BfW_2016_Tanzania_MidTermReport _onlineversion.pdf Upendra Shrestha, "Knowledge for Resilient Livelihoods—Bidhya's Story," *Practical Action* (blog), April 27, 2016, accessed February 14, 2017, http:// practicalaction.org/blog/2016/04/.

42. Bloomberg New Energy Finance and Lighting Global, *Off-Grid Solar Market Trends Report 2016*, February 2016, 1, https://data.bloomberglp.com/bnef/sites/4/2016/03 /20160303_BNEF_WorldBankIFC_Off-GridSolarReport_.pdf.

43. Bloomberg, 2016.

44. REN21, 2018.

45. International Renewable Energy Agency (IRENA), *Innovation Outlook: Renewable Mini-Grids, Summary for Policy Makers* (Abu Dhabi, United Arab Emirates: IRENA, 2016), http://www.irena.org/DocumentDownloads/Publications/IRENA_Innovation _Outlook_Minigrids_Summary_2016.pdf.

46. See figure 4.3.

47. Ernesto Macias Galan, Solarwatt, personal communication with REN21, January 30, 2017.

48. Martin, Starace, and Tricoire, 2017.

49. *The Internet of Risky Things: Presentation to the U.S. Department of Energy Electricity Advisory Comm., Smart Grid Subcomm.* (January 2017) (statement of Bill Sanders); *Internet of Things: Presentation to the U.S. Department of Energy Electricity Advisory Comm., Smart Grid Subcomm.* (March 2017) (statement of Vint Cerf).

50. Alexandra May, *Power to the People: How the Sharing Economy Will Transform the Electricity Industry* (New York: World Economic Forum, 2017), https://www.weforum .org/press/2017/03/power-to-the-people-how-the-sharing-economy-will-transform -the-electricity-industry/.

51. Litos Strategic Communication, *The Smart Grid: An Introduction: How a Smarter Grid Works as an Enabling Engine for Our Economy, Our Environment and Our Future*, U.S. Department of Energy, 2014, accessed October 2017, https://www .energy.gov/sites/prod/files/oeprod/DocumentsandMedia/DOE_SG_Book_Single _Pages%281%29.pdf.

52. European Smart Grid Task Force, *Task Force Smart Grids—Vision and Work Programme*, 2010.

53. This includes two states with voluntary net metering. National Conference of State Legislatures, *State Net Metering Policies*, November 20, 2017, accessed December 19, 2018, http://www.ncsl.org/research/energy/net-metering-policy-overview-and-state -legislative-updates.aspx.

54. D. J. Hess, "The Politics of Niche-Regime Conflicts: Distributed Solar Energy in the United States," *Environmental Innovation and Societal Transitions* 19 (2016): 42–50.

55. Bloomberg New Energy Finance, *Q1 2016 Digital Energy Market Outlook*, February 25, 2016.

56. Juan Pedro Tomas, "10 Million Smart Meters Deployed in Japanese Grid Modernization Project," *Enterprise iot insights*, May 2, 2017, accessed December 18, 2018, https://enterpriseiotinsights.com/20170502/internet-of-things/20170502internet-of -things20170502internet-of-things-japan-tepco-smart-meters-grid-modernization -project-tag23.

57. Marilyn A. Brown, Shan Zhou, and Majid Ahmadi, "Smart-Grid Governance: An International Review of Evolving Policy Issues and Innovations," *Wiley Interdisciplinary Reviews (WIREs): Energy and Environment* 7, no. 5 (2018): e290; M. A. Brown and B. K. Sovacool, *Climate Change and Global Energy Security: Technology and Policy Options* (Cambridge, MA: MIT Press, 2014).

58. State of New York Public Service Commission, *Case 14-M-0101—Proceeding on Motion of the Commission in Regard to Reforming the Energy Vision*, 2015, accessed May 4, 2019, http://documents.dps.ny.gov/public/MatterManagement/CaseMaster .aspx?MatterCaseNo=14-m-0101.

59. Italian Ministry for Economic Development, *Italian National Renewable Energy Action Plan* (in line with the provisions of Directive 2009/28/EC), June 30, 2010, http://www .buildup.eu/sites/default/files/content/national_renewable_energy_action_plan _italy_en.pdf.

60. European Commission, *Operational Programme "Renewable Energy and Energy Efficiency,"* 2007, http://ec.europa.eu/regional_policy/en/atlas/programmes/2007-2013 /italy/operational-programme-renewable-energy-and-energy-efficiency.

61. Italian Ministry for Economic Development, 2010.

62. International Energy Agency (IEA), *Technology Roadmap: Smart Grids* (Paris: IEA, 2011).

63. S. Renner et al., *European Smart Metering Landscape Report* (Vienna: Austrian Energy Agency, 2011), https://www.scribd.com/doc/91236275/D2-1-European-Smart -Metering-Landscape-Report-Final.

64. Renner et al., 2011.

65. Enel Distribuzione, *Smart Meters: Consumption Becomes Smarter and Smarter*, March 11, 2016, https://www.enel.com/media/news/d/2016/03/smart-meters-consumption -becomes-smarter-and-smarter.

66. McDermott Will and Emery, "Italy Issues Fourth Conto Energia: New Feed-In Tariffs for Production of Photovoltaic Energy in 2011–2016," *JDSupra*, May 6, 2011, https:// www.jdsupra.com/legalnews/italy-issues-fourth-conto-energia-new-f-86365/.

67. Smart Energy International, "Organized Crime and EU Solidarity: Enel Italy Talks Cybersecurity," *Metering.com*, November 2, 2015, accessed November 2, 2017, https://www.metering.com/features/enel-italy-talks-cybersecurity/.

68. Presidency of the Council of Ministers, *National Strategic Framework for Cyberspace Security*, December 2013, accessed October 20, 2017, https://www.sicurezzanazionale .gov.it/sisr.nsf/wp-content/uploads/2014/02/italian-national-strategic-framework -for-cyberspace-security.pdf.

69. G. Coraggio, "Updated Guidelines on the European Privacy Regulation by the Italian Data Protection Authority," *GamingTechLaw* (blog), updated May 4, 2018, accessed October 21, 2017, http://www.gamingtechlaw.com/2017/05/privacy-gdpr-italian-data -protection-garante.html.

70. United Nations Environment Program, *Overview of the Republic of Korea's National Strategy for Green Growth*, April 2010, http://www.cdn.giweh.ch/2015/9/21/1-15-24 -1-1.pdf; also refer to the Energy Information Agency's website for more data on South Korea's total primary energy mix: https://www.eia.gov/beta/international /analysis.php?iso=KOR.

71. Taken from Climate Action Tracker, as updated March 10, 2016, accessed March 8, 2018, http://climateactiontracker.org/countries/southkorea/2016.html.

72. Climate Action Tracker, 2016.

73. S. Han, "South Korea Votes to Set Limits on Carbon Emissions," *Bloomberg Businessweek*, August 8, 2012, accessed May 4, 2019, https://www.bloomberg.com /news/articles/2012-02-08/south-korea-moves-closer-to-setting-limits-on-carbon -emissions.

74. H. Bae and C. Wheelock, *Smart Grid in Korea—Smart Power Grid, Consumers, Transportation, Renewable Energy, and Electricity Service: Market Analysis and Forecasts* (Boulder, CO: Pike Research, 2010).

75. Korea Electric Power Corporation (KEPCO), *KEPCO 2011 Annual Report* (Seoul, Korea: KEPCO, 2011), https://www.lacp.com/2010vision/pdf/9444.pdf.

76. Lund, P. D., Byrne, J. A., Haas, R., & Flynn, D. (Eds.). Advances in Energy Systems: The Large-scale Renewable Energy Integration Challenge. (London, UK: Wiley, 2019).

77. Yale Center for Environmental Law & Policy, *Climate Policy & Emissions Data Sheet: South Korea*, 2012.

78. Ministry of Knowledge Economy and Korea Smart Grid Institute, *10 Power IT Projects*, 2005.

79. *South Korea: Smart Grid Revolution*, July 2011, https://www.smartgrid.gov/files/South _Korea_Smart_Grid_Revolution_201112.pdf.

80. Son Jong Cheon, *Smart Grid in Korea*, Korea Smart Grid Institute, 2013, accessed November 6, 2017, http://www.iea.org/media/training/bangkoknov13/session_7c_ksgi _korea_smart_grids.pdf.

81. KEPCO, 2011.

82. Ministry of Knowledge Economy and Korea Smart Grid Institute, 2005.

83. M. Felson and J. L. Spaeth, "Community Structure and Collaborative Consumption: A Routine Activity Approach," *American Behavioral Scientist* 21, no. 4 (1978): 614–624, https://doi.org/10.1177/000276427802100411.

84. C. J. Martin, "The Sharing Economy: A Pathway to Sustainability or a Nightmarish Form of Neoliberal Capitalism?" *Ecological Economics* 121 (2016): 149–159, https:// doi.org/10.1016/j.ecolecon.2015.11.027.

85. M. Cheng, "Sharing Economy: A Review and Agenda for Future Research," *International Journal of Hospitality Management* 57 (2016): 60–70, https://doi.org/10.1016/j. ijhm.2016.06.003.

86. Yochai Benkler, "'Sharing Nicely': On Shareable Goods and the Emergence of Sharing as a Modality of Economic Production," *Yale Law Journal* 114, no. 273 (2004), Faculty Scholarship Series, http://digitalcommons.law.yale.edu/fss_papers/3129/.

87. Russell Belk, "Why Not Share Rather Than Own?" *The Annals of the American Academy of Political and Social Science* 611, no. 1 (2007): 128, doi:10.1177 /0002716206298483.

88. N. A. John, *Sharing, Collaborative Consumption and Web 2.0*, Media@Lse Electronic Working Paper No. 26 (London: Media@Lse, 2013), http://www.lse.ac.uk /media-and-communications/assets/documents/research/working-paper-series /EWP26.pdf.

89. H. Heinrichs, "Sharing Economy: A Potential New Pathway to Sustainability," *GAIA* 22, no. 4 (2013): 228–231.

90. John, 2013.

91. Martin, 2016.

92. K. Finley, *Trust in the Sharing Economy: An Exploratory Study*, University of Warwick, 2013, https://warwick.ac.uk/fac/arts/theatre_s/cp/research/publications/madiss/ccps _a4_ma_gmc_kf_3.pdf.

93. B. Bhushan and S. K. Soonee, *Utilization of Idling Capacity of Captive Power Plants: A Viable Alternative*, https://slideplayer.com/slide/9936564/.

94. D. Demailly and A.-S. Novel, "The Sharing Economy: Make It Sustainable," *Iddri* 3, no. 14 (2014): 30, http://www.phibetaiota.net/wp-content/uploads/2014/10/The -Sharing-Economy-and-Sustainability.pdf.

95. Demailly and Novel, 2014.

96. National Academies of Sciences, Engineering, and Medicine, *Pathways to Urban Sustainability: Challenges and Opportunities for the United States* (Washington, DC: National Academies Press, 2016), appendix A, https://www.eenews.net/assets /2016/10/19/document_gw_04.pdf.

97. J. Orsi, "The Sharing Economy Just Got Real," *Shareable*, September 16, 2013, accessed April 20, 2015, http://www.shareable.net/blog/the-sharingeconomy-just-got-real.

98. G. Zervas, D. Proserpio, and J. Byers, *A First Look at Online Reputation on Airbnb, Where Every Stay Is Above Average*, January 28, 2015, https://ssrn.com/abstract =2554500.

99. L. Gansky, *The Mesh: Why the Future of Business Is Sharing* (New York: Penguin, 2010).

100. Zervas, Proserpio, and Byers, 2015.

101. A. Daunorienė et al., "Evaluating Sustainability of Sharing Economy Business Models," *Procedia—Social and Behavioral Sciences* 213 (2015): 836–841, https://doi .org/10.1016/j.sbspro.2015.11.486.

102. May, 2017.

103. May, 2017.

104. Cheng, 2016.

105. Cheng, 2016.

5. Fostering and Financing the Energy Infrastructure Transition

1. Francis Altdorfer, *Impact of the Economic Crisis on the EU's Industrial Energy Consumption* (policy brief), Odyssee-Mure, April 2017, http://www.odyssee-mure.eu/publications/policy-brief/impact-economic-crisis-industrial-energy-consumption.pdf.

2. Marilyn A. Brown et al., "Exploring the Impact of Energy Efficiency as a Carbon Mitigation Strategy in the U.S.," *Energy Policy* 109 (2017): 249–259.

3. Brown et al., 2017.

4. J. Taylor and P. Van Doren, "Energy Myth Five: Price Signals Are Insufficient to Induce Efficient Energy Investments," in *Energy and American Society: Thirteen Myths*, ed. B. K. Sovacool and M. A. Brown (Dordrecht, Netherlands: Springer, 2007), 125–144.

5. L. J. Makovich, *The Cost of Energy Efficiency Investments: The Leading Edge of Carbon Abatement* (Cambridge: IHS CERA, 2008), 15.

6. S. Gupta et al., "Cross-cutting Investment and Finance Issues," in IPCC, 2014: *Climate Change 2014: Mitigation of Climate Change. Contribution of Working Group III to the Fifth Assessment Report of the Intergovernmental Panel on Climate Change*, ed. O. Edenhofer et al. (Cambridge: Cambridge University Press, 2014).

7. National Academy of Sciences (NAS), *Limiting the Magnitude of Future Climate Change* (Washington, DC: NAS, 2010).

8. M. A. Brown, "Market Failures and Barriers as a Basis for Clean Energy Policies," *Energy Policy* 29 (2001): 1197–1207.

9. Matthew Antes et al., *Workshops on R&D Opportunities in Clean Energy Innovation: A How-To Guide for Mission Innovation Members* (Oak Ridge, TN: Oak Ridge National Laboratory, 2017), http://mission-innovation.net/wp-content/uploads/2017/12/MI-Workshop-Guide-2017-October-Final.pdf.

10. Daniel L. Sanchez and Varun Sivaram, "Saving Innovative Climate and Energy Research: Four Recommendations for Mission Innovation," *Energy Research & Social Science* 29 (2017): 123–126.

11. http://mission-innovation.net/2016/06/02/inaugural-mission-innovation-ministerial-pledges-unprecedented-support-for-clean-energy-research-and-development/.

12. Z. Myslikova, K. S. Gallagher, and F. Zhang, *Mission Innovation 2.0: Recommendations for the Second Mission Innovation Ministerial in Beijing, China*, CIERP Climate Policy Lab Discussion Paper, Tufts University, 2017, https://sites.tufts.edu/cierp/files/2017/09/CPL_MissionInnovation014_052317v2low.pdf.

13. Antes et al., 2017.

14. Lu Wang, Yi-Ming Wei, and M. A. Brown, "Global Transition to Low-Carbon Electricity: A Bibliometric Analysis," *Applied Energy* 205 (2017): 57–68, https://doi.org/10.1016/j.apenergy.2017.07.107.

15. David Danielson, "Breakthrough Energy Ventures" (presentation to the Georgia Institute of Technology, December 15, 2017).

16. D. M. Kammen and G. F. Nemet, "Energy Myth Eleven: R&D Investment Takes Decades to Reach the Market," in Sovacool and Brown, 2007.

17. Wang, Wei, and Brown, 2017.

18. Organization for Economic Cooperation and Development (OECD), *Research Co-operation Between Developed and Developing Countries in the Area of Climate Change Adaptation and Biodiversity*, 2014, http://www.oecd.org/science/Research _Cooperation.pdf.

19. Wang, Wei, and Brown, 2017.

20. Q. Hou et al., "Mapping the Scientific Research on Life Cycle Assessment: A Biblio-metric Analysis, *International Journal of Life Cycle Assessment* 20 (2015): 541–555; I. Mignon and A. Bergek, "Investments in Renewable Electricity Production: The Importance of Policy Revisited," *Renewable Energy* 88 (2016): 307–316; W. White et al., "The Role of Governments in Renewable Energy: The Importance of Policy Consistency," *Biomass and Bioenergy* 57 (2013): 97–105.

21. M. Brown and Y. Wang, *Green Savings: How Policies and Markets Drive Energy Efficiency* (Santa Barbara, CA: Praeger, 2015).

22. E. Massetti, "The Macroeconomics of Climate Policy: Investments and Financial Flows," in *Towards a Workable and Effective Climate Regime*, ed. Scott Barrett, Carlo Carraro, and Jaime de Melo (London: Center for Economic Policy Research Press, 2015), 467–482, http://voxeu.org/sites/default/files/file/massetti.pdf; G. C. Iyer et al., "Improved Representation of Investment Decisions in Assessments of CO_2 Mitigation," *Nature Climate Change* 5 (2015): 436–440.

23. Gupta et al., 2014.

24. Gupta et al., 2014, 1211.

25. Data accessed December 28, 2017, at https://www.co2.earth/global-co2-emissions.

26. Gupta et al., 2014.

27. Renewable Energy Policy Network for the 21st Century (REN21), *Renewables 2018 Global Status Report*, http://www.ren21.net/gsr-2018/.

28. Interview with Tim Rockell, director of Global Energy Institute, KPMG, and Sharad Somani, partner in Global Infrastructure Advisory, KPMG, "Energy Financing Challenges: Goal, Gas, and Renewables," *DNV-GL*, 2019, https://www.dnvgl.com /energy/publications/podcast/pc-energy-financing.html.

29. Interview with Rockell and Somani, 2019.

30. Interview with Rockell and Somani, 2019.

31. Gupta et al., 2014.

32. C. Carraro, A. Favero, and E. Massetti, "Investments and Public Finance in a Green, Low Carbon, Economy," *Energy Economics* 34 (2012): S15–S28; V. Chaturvedi et al., "Capital Investment Requirements for Greenhouse Gas Emissions Mitigation in Power Generation on Near Term to Century Time Scales and Global to Regional Spatial Scales," *Energy Economics* 46 (2014): 267–278; T. Kober et al., "A Multi-Model Study of Energy Supply Investments in Latin America Under Climate Control Policy," *Energy Economics* 56 (2016): 543–551.

33. Marilyn A. Brown et al., 2017.

34. Marilyn A. Brown and Yufei Li, "Carbon Pricing and Energy Efficiency: Pathways to Deep Decarbonization of the U.S. Electric Sector," *Energy Efficiency* 12, no. 2 (2019): 463–481.

35. As of 2018, the Trump administration has directed the EPA to examine strategies for repealing the Clean Power Plan. For more on this, refer to https://grist.org/article /trump-epa-wants-to-repeal-clean-power-plan-even-though-it-would-save-thousands -of-lives/.

36. REN21, 2018.

37. REN21, 2018.

38. S. V. Valentine, "The Socio-Political Economy of Electricity Generation in China," *Renewable and Sustainable Energy Reviews* 32 (2014): 416–429.

39. M. Xu et al., "Gigaton Problems Need Gigaton Solutions," *Environmental Science & Technology* 44, no. 11 (2010): 4037–4041, http://dx.doi.org/10.1021/es903306e.

40. https://www.tva.gov/Newsroom/Press-Releases/TVA-Board-Approves-300 -Million-Strategic-Fiber-Initiative.

41. National Academies of Sciences, Engineering, and Medicine, *Pathways to Urban Sustainability: Challenges and Opportunities for the United States*, 2016, http://sites .nationalacademies.org/PGA/sustainability/urbanstudy/index.htm.

42. World Bank and Ecofys, *Carbon Pricing Watch 2017* (Washington, DC: World Bank, 2017), https://openknowledge.worldbank.org/handle/10986/26565.

43. World Bank and Ecofys, 2017.

44. World Bank and Ecofys, 2017.

45. Gupta et al., 2014, 1216.

46. John Horowitz et al., *Methodology for Analyzing a Carbon Tax*, Working Paper 115 (Washington, DC: Office of Tax Analysis, 2017).

47. C. A. Grainger and C. D. Kolstad, "Who Pays a Price on Carbon?" *Environmental and Resource Economics* 46, no. 3 (2010): 359–376; A. Chamberlain, *Who Pays for Climate Policy? New Estimates of the Household Burden and Economic Impact of a U.S. Cap-and-Trade System* (Washington, DC: Tax Foundation, 2009), http://www .taxfoundation.org/files/wp6.pdf; M. Shammin and C. Bullard, "Impact of Cap-and -Trade Policies for Reducing Greenhouse Gas Emissions on U.S. Households," *Ecological Economics* 68, nos. 8–9 (2009): 2432–2438.

48. W. J. Baumol and W. E. Oates, *The Theory of Environmental Policy*, 2nd ed. (Cambridge: Cambridge University Press, 1988).

49. Richard Tol, "The Structure of the Climate Debate," *Energy Policy* 104 (2017): 431–438.

50. Tol, 2017.

51. Peak-load capacity is the amount of installed capacity of energy systems that are capable of responding in a timely manner to supply and demand fluctuations. Common peak-load technologies include hydropower and natural gas–fired power plants.

52. L. Goulder and I. W. H. Parry, "Instrument Choice in Environmental Policy," *Review of Environmental Economics and Policy* 2 (2008): 152–174.

53. Paul Hawken, *Drawdown: The Most Comprehensive Plan Ever Proposed to Reverse Global Warming* (New York: Penguin Books, 2017).

54. D. Arent et al., "Key Economic Sectors and Services," in IPCC, 2014: *Climate Change 2014: Impacts, Adaptation, and Vulnerability. Part A: Global and Sectoral Aspects*, ed. C. B. Field et al. (Cambridge: Cambridge University Press, 2014), 659–708; Richard S. J. Tol, "The Economic Impact of Climate Change in the 20th and 21st Centuries," *Climatic Change* 117, no. 4 (2013): 795–808.

55. Q. M. Liang and Y. M. Wei, "Distributional Impacts of Taxing Carbon in China: Results from the CEEPA Model," *Applied Energy* 92 (2012): 545–551, doi:10.1016/j. apenergy.2011.10.036; J. Dodson and N. Sipe, "Oil Vulnerability in the Australian City: Assessing Socioeconomic Risks from Higher Urban Fuel Prices," *Urban Studies* 44, no. 1 (2007): 37–62.

56. S. Carley, "The Era of State Energy Policy Innovation: A Review of Policy Instruments," *Review of Policy Research* 28, no. 3 (2011): 265–294.

57. D. Burtraw, R. Sweeney, and M. Walls, "The Incidence of U.S. Climate Policy: Where You Stand Depends on Where You Sit," *SSRN*, RRF Discussion Paper No. 08-28, September 15, 2008, https://papers.ssrn.com/sol3/papers.cfm?abstract_id=1272667.

58. H. Fell and R. D. Morgenstern, "Alternative Approaches to Cost Containment in a Cap-and-Trade System" (working paper), *Resources for the Future*, April 13, 2009, https://www.rff.org/publications/working-papers/alternative-approaches-to-cost -containment-in-a-cap-and-trade-system/.

59. Marc Chesney et al., *Environmental Finance and Investments*, 2nd ed. (Heidelberg, Germany: Springer, 2016).

60. M. Beck et al., "Carbon Tax and Revenue Recycling: Impacts on Households in British Columbia," *Resource and Energy Economics* 41 (2015): 40–69, https://doi .org/10.1016/j.reseneeco.2015.04.005; T. Callan et al., "The Distributional Implications of a Carbon Tax in Ireland," *Energy Policy* 37 (2009): 407–412, https://doi.org/10.1016/j .enpol.2008.08.034; Y. Liu and Y. Lu, "The Economic Impact of Different Carbon Tax Revenue Recycling Schemes in China: A Model-Based Scenario Analysis," *Applied Energy* 141 (2015): 96–105, https://doi.org/10.1016/j.apenergy.2014.12.032; B. Murray and N. Rivers, "British Columbia's Revenue-Neutral Carbon Tax: A Review of the Latest Grand Experiment in Environmental Policy," *Energy Policy* 86 (2015): 674–683, https://doi.org/10.1016/j.enpol.2015.08.011.

61. "Paris Agreement Ratification Tracker," *Climate Analytics*, accessed July 17, 2018, https://climateanalytics.org/briefings/ratification-tracker/.

62. World Bank and Ecofys, 2017.

63. World Bank and Ecofys, 2017.

64. The World Bank publishes the Carbon Pricing Dashboard website, adding an interactive dimension to the annual State and Trends of Carbon Pricing reports. This resource provides an up-to-date overview of carbon pricing initiatives and allows users to navigate through the visuals and data of *Carbon Pricing Watch 2017*. It can be accessed via http://carbonpricingdashboard.

65. World Bank and Ecofys, 2017.

66. CDP (formerly the Carbon Disclosure Project), *Embedding a Carbon Price into Business Strategy*, September 2016, accessed December 29, 2017, https://

b8f65cb373b1b7b15feb-c70d8ead6ced550b4d987d7c03fcdd1d.ssl.cf3.rackcdn.com /cms/reports/documents/000/001/132/original/CDP_Carbon_Price_report_2016 .pdf?1474899276.

67. UNEP Finance Initiative, *Task Force on Climate-Related Financial Disclosures*, accessed December 18, 2018, http://www.unepfi.org/climate-change/tcfd/.

68. World Bank and Ecofys, 2017.

69. M. A. Brown and B. K. Sovacool, *Climate Change and Global Energy Security: Technology and Policy Options* (Cambridge, MA: MIT Press, 2014).

70. Brown and Wang, 2015.

71. Database of State Incentives for Renewables & Efficiency (DSIRE), http://www.dsireusa .org/.

72. K. Neuhoff et al., *Financial Incentives for Energy Efficiency Retrofits in Buildings*, Proceedings of the ACEEE Summer Study on Energy Efficiency in Buildings, 2012, https://aceee.org/files/proceedings/2012/data/papers/0193-000422.pdf.

73. G. Valentini and P. Pistochini, "The 55 Percent Tax Reductions for Building Retrofitting in Italy: The Results of the ENEA's Four Years Activities," *6th EEDAL Conference*, 2011, http:// openarchive.enea.it/bitstream/handle/10840/4813/DEF_161_ENEA_Gianpaolo %20Valentini_Patrizia%20Pistochini_final_paper.pdf?sequence=1, accessed May 3, 2019.

74. International Energy Agency (IEA), *Energy Efficiency Market Report 2013* (Paris: IEA, 2013), https://www.iea.org/publications/freepublications/publication/EEMR2013 _free.pdf.

75. Brown and Wang, 2015.

76. C. J. Bell, S. Nadel, and S. Hayes, *On-Bill Financing for Energy Efficiency Improvements* (Washington, DC: American Council for an Energy-Efficient Economy, 2011), http:// www.puc.state.pa.us/Electric/pdf/Act129/OBF-ACEEE_OBF_EE_Improvements .pdf.

77. Bell, Nadel, and Hayes, 2011.

78. N. Zobler and K. Hatcher, "Financing Energy Efficiency Projects," *Government Finance Review* 19 (2003): 14–18.

79. G. Kats, *Greening Our Built World: Costs, Benefits, and Strategies* (Washington, DC: Island Press, 2010); A. Marino et al., "A Snapshot of the European Energy Service Market in 2010 and Policy Recommendations to Foster a Further Market Development," *Energy Policy* 39 (2011): 6190–6198.

80. L. Schewel et al., *Top Federal Energy Policy Goals*, Rocky Mountain Institute, 2009, accessed May 3, 2019, https://rmi.org/insight/rmis-top-federal-energy-policy-goals/.

81. D. J. Hess, "The Politics of Niche-Regime Conflicts: Distributed Solar Energy in the United States," *Environmental Innovation and Societal Transitions* 19 (2016): 42–50.

82. R. C. Headen et al., "Property Assessed Clean Energy Financing: The Ohio Story," *The Electricity Journal* 24 (2011): 47–56.

83. DSIRE, 2012.

84. http://www.renewableenergyworld.com/rea/news/article/2011/11/feed-in-tariffs-best -to-deal-with-climate-change-says-ipcc-working-group-iii-renewables.

85. REN21, 2018.

86. L. M. MacLean and J. N. Brass, "Foreign Aid, NGOs and the Private Sector: New Forms of Hybridity in Renewable Energy Provision in Kenya and Uganda," *Africa Today* 62, no. 1 (2015): 56–82.

87. F. Lambe et al., "Can Carbon Finance Transform Household Energy Markets? A Review of Cookstove Projects and Programs in Kenya," *Energy Research & Social Science* 5 (2015): 55–66.

88. H. Zerriffi, "Innovative Business Models for the Scale-up of Energy Access Efforts for the Poorest," *Current Opinion in Environmental Sustainability* 3, no. 4 (2011): 272–278.

89. B. K. Sovacool and I. M. Drupady, "Summoning Earth and Fire: The Energy Development Implications of Grameen Shakti in Bangladesh," *Energy* 36, no. 7 (July 2011): 4445–4459.

90. Lambe et al., 2015.

91. J. M. Fontaine, C. Dargnies, and A. Delalande, "Developing Social Business Initiatives for Access to Energy, a Key Success Factor for Sustainability: The Awango Project" (paper presented at the SPE International Conference and Exhibition on Health, Safety, Security, Environment, and Social Responsibility, Stavanger, Norway, April 11–13, 2016).

92. Lighting a Billion Lives (LaBL), *Evolution*, The Energy and Resources Institute, 2015, http://labl.teriin.org/evolution.php.

93. More on this program at http://labl.teriin.org/.

94. Lighting a Billion Lives (LaBL), *Delivery Model*, The Energy and Resources Institute, 2015, http://labl.teriin.org/delivery_model.php.

95. D. R. Limaye and E. S. Limaye, "Scaling Up Energy Efficiency: The Case for a Super ESCO," *Energy Efficiency* 4 (2011): 133–144, https://www.academia.edu/29880438 /Scaling_up_energy_efficiency_the_case_for_a_Super_ESCO.

96. Da-li Gan, "Energy Service Companies to Improve Energy Efficiency in China: Barriers and Removal Measures," *Procedia Earth and Planetary Science* 1 (2009): 1695–1704.

97. Limaye and Limaye, 2011; S. Singh et al., *Energy Efficiency Financing Option Papers for Georgia* (Washington, DC: World Bank Group, 2016), http://documents .worldbank.org/curated/en/825761475845097689/Energy-efficiency-financing -option-papers-for-Georgia.

98. http://production.presstogo.com/fileroot7/gallery/DNVGL/files/original /129779578f7544798859f8840d4afab999.pdf.

6. Policies for Driving Innovation and Expediting the Transition

1. Thomas Friedman, "The Power of Green," *New York Times Magazine*, April 15, 2007, http://www.nytimes.com/2007/04/15/magazine/15green.t.html.

2. Peter Christoff, "Cold Climate in Copenhagen: China and the United States at COP15," *Environmental Politics* 19, no. 4 (2010): 646.

3. S. V. Valentine, "The Socio-Political Economy of Electricity Generation in China," *Renewable and Sustainable Energy Reviews* 32 (2014): 416–429.

4. For data, refer to the Global Wind Energy website at http://gwec.net/wp-content /uploads/2018/04/5_Top-10-cumulative-capacity-Dec-2017-1.jpg, accessed May 3, 2019.

5. Keith Bradsher, "China Looks to Capitalize on Clean Energy as U.S. Retreats," *New York Times*, June 5, 2017, https://www.nytimes.com/2017/06/05/business/energy-environment/china-clean-energy-coal-pollution.html.

6. Ourenergypolicy.org, Department of Energy Notice of Proposed Rulemaking Grid Resiliency Pricing Rule, September 2017, accessed December 30, 2017, http:// www.ourenergypolicy.org/department-of-energy-notice-of-proposed-rulemaking -grid-resiliency-pricing-rule/.

7. Kat Kerlin, "Alternative Fuels Need More Than Hype to Drive Transportation Market," *UCDavis*, March 2, 2016, https://www.ucdavis.edu/news/alternative-fuels-need -more-hype-drive-transportation-market/.

8. Hunt Allcott and Michael Greenstone, "Is There an Energy Efficiency Gap?" *Journal of Economic Perspectives* 26 (2012): 3–28; J. Taylor, "Energy Conservation and Efficiency: The Case Against Coercion," *Policy Analysis* 189 (1993): 1–13.

9. M. A. Brown, "Market Failures and Barriers as a Basis for Clean Energy Policies," *Energy Policy* 29 (2001): 1197–1207.

10. David Weimer and Aidan R. Vining, *Policy Analysis: Concepts and Practice*, 5th ed. (Abingdon, Oxon: Routledge, 2016); K. Gillingham, R. Newell, and K. Palmer, "Energy Efficiency Economics and Policy," *Resources for the Future* (Washington, DC: Resources for the Future, 2009).

11. N. Oreskes and E. M. Conway, *Merchants of Doubt: How a Handful of Scientists Obscured the Truth on Issues from Tobacco Smoke to Global Warming* (New York: Bloomsbury Publishing, 2011).

12. H. Tabuchi, "Uranium Miners Pushed Hard for a Comeback. They Got Their Wish," *New York Times*, January 13, 2018, https://www.nytimes.com/2018/01/13/climate /trump-uranium-bears-ears.html.

13. C. C. Hood and H. Z. Margetts, *The Tools of Government in the Digital Age* (New York: Palgrave Macmillan, 2007).

14. Michele Gorman, "Yogi Berra's Most Memorable Sayings," *Newsweek*, September 23, 2015, accessed December 30, 2017, http://www.newsweek.com/most-memorable -yogi-isms-375661.

15. C. Hood, *The Tools of Government* (Chatham, NJ: Chatham House, 1986).

16. M. Howlett and M. Ramesh, *Studying Public Policy: Policy Cycles and Policy Subsystems* (Don Mills, Ontario: Oxford University Press, 1995).

17. Marilyn A. Brown, Rodrigo Cortes-Lobos, and Matthew Cox, "Reinventing Industrial Energy Use in a Resource-Constrained World," in *Energy, Sustainability and the Environment*, ed. Fereidoon Sioshansi (Burlington, MA: Elsevier Press, 2011), chapter 12, 337–366.

18. L. Harrington, J. Brown, and M. Caithness, *Energy Standards and Labelling Programs Throughout the World in 2013*, report prepared for Energy Efficient Strategies and Maia Consulting for the Australian Department of Industry, 2014.

19. https://ec.europa.eu/info/energy-climate-change-environment/standards-tools -and-labels/products-labelling-rules-and-requirements/energy-label-and -ecodesign_en.

20. http://ec.europa.eu/environment/ecolabel.

21. Marilyn A. Brown, "Innovative Energy-Efficiency Policies: An International Review," *Wiley Interdisciplinary Reviews (WIREs): Energy and Environment* 4, no. 1 (2015): 1–25, https://onlinelibrary.wiley.com/doi/abs/10.1002/wene.125.

22. International Energy Agency (IEA), *Energy Efficiency Market Report 2013* (Paris, IEA, 2013).

23. More on this program can be found at http://www.energylabel.gov.cn/ (only available in Mandarin).

24. M. Ellis, I. Barnsley, and S. Holt, *Barriers to Maximizing Compliance with Energy Efficiency Policy* (La Colle sur Loup, France: European Council for an Energy Efficient Economy, 2009), 341–351.

25. Ellis, Barnsley, and Holt, 2009.

26. A. Goett, *Household Appliance Choice: Revision of REEPS Behavioral Models* (Cambridge Systematics, Palo Alto, CA, 1983); M. Coller and M. Williams, "Eliciting Individual Discount Rates," *Experimental Economics* 2 (1999): 107–127.

27. C. Robinson et al., "Machine Learning Approaches for Estimating Commercial Building Energy Consumption," *Applied Energy* 208 (2017): 889–904, https://www .sciencedirect.com/science/article/pii/S0306261917313429.

28. John Lee et al., *plaNYC: New York City Local Law 84 Benchmarking Report* (New York: Office of the Mayor, 2014); New York City Mayor's Office of Sustainability, "Local Law 84 Data Disclosures," 2016, accessed May 4, 2019, https://www1.nyc.gov/html /gbee/html/plan/ll84.shtml.

29. Lee et al., 2014.

30. J. Christmas, "Financing Energy Efficiency in the Commercial Building Sector: Is There Hope Post-PACE?" (paper presented at the Fifth Annual Energy Efficiency Finance Forum, Philadelphia, 2011), http://www.aceee.org/files/pdf/conferences /eeff/2011/2011presentations.pdf; I. A. Campbell, "Tapping into a Trillion Dollar Industry: How to Increase Energy Efficiency Financing by 2015" (paper presented at the Fifth Annual Energy Efficiency Finance Forum, Philadelphia, 2011, http:// www.aceee.org/files/pdf/conferences/eeff/2011/2011presentations.pdf;N. Miller, J. Spivey, and A. Florance, *Does Green Pay Off?* 2008, https://www.energystar.gov /sites/default/files/buildings/tools/DoesGreenPayOff.pdf; J. Jackson, "How Risky Are Sustainable Real Estate Projects? An Evaluation of LEED and Energy Star Development Options," *Journal of Sustainable Real Estate* 1 (2009): 91–106; P. Das, A. Tidwell, and A. Ziobrowski, "Dynamics of Green Rentals over Market Cycles: Evidence from Commercial Office Properties in San Francisco and Washington DC," *Journal of Sustainable Real Estate* 3 (2011): 1–22.

31. Matt Cox, Marilyn A. Brown, and Xiaojing Sun, "Energy Benchmarking of Commercial Buildings: A Low-Cost Pathway for Urban Sustainability," *Environmental Research Letters* 8, no. 3 (2013): 035018, http://iopscience.iop.org/1748-9326/8/3/035018.

32. http://pacenation.us/pace-in-georgia/.

33. D. Saygin et al., "Benchmarking the Energy Use of Energy-Intensive Industries in Industrialized and in Developing Countries," *Energy* 36 (2011): 6661–6673; D. Phylipsen et al., "Benchmarking the Energy Efficiency of Dutch Industry: An Assessment of the Expected Effect on Energy Consumption and CO_2 Emissions," *Energy Policy* 8 (2002): 663–679; W. Nuijen and M. Booij, "Experiences with Long Term Agreements on Energy Efficiency and an Outlook to Policy for the Next 10 Years," NOVEM: Ultrech, NL, 2002.

34. N. Jollands et al., "The 25 IEA Energy Efficiency Policy Recommendations to the G8 Gleneagles Plan of Action," *Energy Policy* 38 (2010): 6409–6418; International Energy Agency (IEA), *Energy Security and Climate Policy: Assessing Interactions* (Paris: IEA, 2007).

35. L. Price and A. T. McKane, *Policies and Measures to Realize Industrial Energy Efficiency and Mitigate Climate Change*, report prepared by UN-Energy, 2009, http:// www.unido.org/fileadmin/user_media/Services/Energy_and_Climate_Change /EPU/UN Energy 2009 and Measures to realise%Industrial Energy Efficiency and mitigate Climate_small.pdf.

36. S. Rezessy and P. Bertoldi, "Voluntary Agreements in the Field of Energy Efficiency and Emission Reduction: Review and Analysis of Experiences in the European Union," *Energy Policy* 39 (2011): 7121–7129.

37. Building and Construction Authority, *BCA Building Energy Benchmarking Report 2018*, accessed December 18, 2018, https://www.bca.gov.sg/GreenMark/others/BCA _BEBR_Abridged_FA_2018.pdf.

38. Brown, M. A., & Wang, Y. (2015). *Green savings: how policies and markets drive energy efficiency: how policies and markets drive energy efficiency*. ABC-CLIO, 98

39. S. Curkovic, R. Sroufe, and S. Melnyk, "Identifying the Factors Which Affect the Decision to Attain ISO 14000," *Energy* 30, no. 8 (2005): 1387–1407.

40. Daniel C. Matisoff, Douglas S. Noonan, and John J. O'Brien, "Convergence in Environmental Reporting: Assessing the Carbon Disclosure Project," *Business Strategy and the Environment* 22, no. 5 (2013): 285–305.

41. M. Piecyk, "Carbon Auditing of Companies, Supply Chains, and Products," in *Green Logistics: Improving the Environmental Sustainability of Logistics*, ed. A. C. McKinnon et al. (London: Kogan Page Limited, 2010).

42. S. V. Valentine, "The Green Onion: A Corporate Environmental Strategy Framework," *Corporate Social Responsibility and Environmental Management* 17, no. 5 (2010): 284–298.

43. A. McKane, *Industrial Energy Management: Issues Paper*, Lawrence Berkeley National Laboratory, 2007, http://www.unido.org/fieladmin/import/63563_EM_Issues_Paper031207 .pdf.

44. Intergovernmental Panel on Climate Change (IPCC), 2014: *Climate Change 2014: Mitigation of Climate Change. Contribution of Working Group III to the Fifth Assessment Report of the Intergovernmental Panel on Climate Change*, ed. O. Edenhofer et al. (Cambridge: Cambridge University Press, 2014).

45. More on NEA energy efficiency incentives can be found at http://www.e2singapore .gov.sg/Incentives/Guide.aspx.

46. T. Fleiter et al., "Energy Efficiency in the German Pulp and Paper Industry: A Model -Based Assessment of Saving Potentials," *Energy* 40 (2012): 84–99.

47. E. Gruber et al., "Efficiency of an Energy Audit Programme for SMEs in Germany: Results of an Evaluation Study," *ECEEE 2011 Summer Study, Energy Efficiency First: The Foundation of a Low-Carbon Society*, Vol. 2: Panel 3 (2011), 663–673, https://www .eceee.org/library/conference_proceedings/eceee_Summer_Studies/2011/3-energy -use-in-industry-the-road-from-policy-to-action/efficiency-of-an-energy-audit -programme-for-smes-in-germany-results-of-an-evaluation-study/.

48. O. Kimura, *Japanese Top Runner Approach for Energy Efficiency Standards*, CRIEPI, 2010, http://criepi.denken.or.jp/en/serc/research_re/download/09035dp.pdf.

49. For more on this program, see http://www.energy.gov/eere/amo/industrial-assessment -centers-iacs.

50. Marilyn A. Brown et al., "Evaluating the Risks of Alternative Energy Policies: A Case Study of Industrial Energy Efficiency," *Energy Efficiency* 7, no. 1 (2014): 1–22.

51. M.-L. Bemelmans-Videc, R. Rist, and E. Vedung, eds., *Carrots, Sticks, and Sermons: Policy Instruments and Their Evaluation* (Abingdon, Oxon: Routledge, 2017).

52. L. H. Goulder and I. W. Parry, "Instrument Choice in Environmental Policy," *Review of Environmental Economics and Policy* 2, no. 2 (2008): 152–174.

53. Howlett and Ramesh, 1995.

54. P. Lanoie and E. Al, "Environmental Policy, Innovation and Performance: New Insights on the Porter Hypothesis," *Journal of Economics and Management Strategy* 20, no. 3 (2011): 803–842; *Science Advisory Board (SAB) Advisory on EPA's Draft Guidelines for Preparing Economic Analyses*, 2009, https://yosemite.epa.gov/sab /sabproduct.nsf/559B838F18C36F078525763C0058B32F/$File/EPA-SAB-09-018 -unsigned.pdf; S. Ambec et al., "The Porter Hypothesis at 20: Can Environmental Regulation Enhance Innovation and Competitiveness?" *Review of Environmental Economics and Policy* 7, no. 1 (2013): 2–22; Y. Rubashkina, M. Galeotti, and E. Ver-dolini, "Environmental Regulation and Competitiveness: Empirical Evidence on the Porter Hypothesis from European Manufacturing Sectors," *Energy Policy* 83 (2015): 288–300.

55. D. Popp, "International Innovation and Diffusion of Air Pollution Control Tech-nologies: The Effects of NOX and SO_2 Regulation in the US, Japan, and Germany," *Journal of Environmental Economics and Management* 51, no. 1 (2006): 46–71.

56. Lanoie and Al, 2011.

57. https://www.environment.gov.au/climate-change/renewable-energy-target-scheme.

58. http://ec.europa.eu/energy/en/topics/energy-strategy/2050-energy-strategy.

59. https://climateactiontracker.org/countries/india/, accessed May 4, 2019.

60. In Mandarin, http://www.sdpc.gov.cn/zcfb/zcfbghwb/201612/P020161222570036010274 .pdf.

61. These policies are described at http://www.ncsl.org/research/energy/renewable -portfolio-standards.aspx.

62. For more on this subject, see http://www.ncsl.org/research/energy/renewable-port-folio-standards.aspx.

63. S. Valentine, "Japanese Wind Energy Development Policy: Grand Plan or Group Think?" *Energy Policy* 39, no. 11 (2011): 6842–6854.

64. More on this policy at the EIA website: https://www.iea.org/policiesandmeasures /pams/belgium/name-167352-en.php, accessed January 8, 2018.

65. For more on these programs, see the United Nations Climate Change site at https:// unfccc.int/process/the-kyoto-protocol/mechanisms, accessed May 4, 2019.

66. For more on these programs, see the EU ETS site at https://ec.europa.eu/clima/policies /ets_en, accessed January 8, 2018.

67. International Carbon Action Partnership (ICAP), *Emissions Trading Worldwide: Executive Summary*, ICAP Status Report 2017 (Berlin: ICAP, 2017), https://icapcarbonaction .com/en/?option=com_attach&task=download&id=436, accessed January 8, 2018.

68. National Research Council, *Limiting the Magnitude of Future Climate Change*, Consensus Study Report (Washington, DC: The National Academies Press, 2010), https://doi.org/10.17226/12785.

69. L.-B. Desroches et al., "Appliance Standards and Advanced Technologies," *AIP Conference Proceedings* 1401, no. 1 (2011): 339–352, doi:10.1063/1.3653862.

70. U.S. Department of Energy, "About the Appliance and Equipment Standards Program," *Energy.gov*, https://www.energy.gov/eere/buildings/about-appliance-and-equipment -standards-program, accessed December 18, 2018.

71. R. Komiyama and C. Marnay, "Japan's Residential Energy Demand Outlook to 2030 Considering Energy Efficiency Standards 'Top-Runner Approach,'" *2008 ACEEE Summer Study on Energy Efficiency in Buildings*, 2008, https://emp.lbl.gov/publi-cations/japan-s-residential-energy-demand; P. J. S. Siderius and H. Nakagami, "A MEPS Is a MEPS Is a MEPS: Comparing Ecodesign and Top Runner Schemes for Setting Product Efficiency Standards," *Energy Efficiency* 6 (2012): 1–19.

72. Jollands et al., 2010.

73. M. Thiruchelvam and S. Kumar, "Policy Options to Promote Energy Efficient and Environmentally Sound Technologies in Small- and Medium-Scale Industries," *Energy Policy* 31 (2003): 977–987.

74. Desroches et al., 2011.

75. N. M. Sachs, "Can We Regulate Our Way to Energy Efficiency? Product Standards as Climate Policy," *Vanderbilt Law Review* 65 (2012): 1631–1678.

76. H. C. Granade et al., *Unlocking Energy Efficiency in the U.S. Economy*, McKinsey and Company, New York, 2009, https://www.mckinsey.com/~/media/mckinsey/dotcom /client_service/epng/pdfs/unlocking%20energy%20efficiency/us_energy_efficiency _exc_summary.ashx.

77. A. T. Chan and V. C. H. Yeung, "Implementing Building Energy Codes in Hong Kong: Energy Savings, Environmental Impacts and Cost," *Energy and Buildings* 37 (2005): 631–642; X. Sun et al., *Making Buildings Part of the Climate Solution by Enforcing Aggressive Commercial Building Codes*, Georgia Institute of Technology, 2012, https://smartech.gatech.edu/handle/1853/45599.

78. J. Weiss, E. Dunkelberg, and T. Vogelpohl, "Improving Policy Instruments to Better Tap into Homeowner Refurbishment Potential: Lessons Learned from a Case Study in Germany," *Energy Policy* 44 (2012): 406–415.

79. The European Parliament and the Council of the European Union, "Directive 2010/31 /EU of the European Parliament and of the Council of 19 May 2010 on the Energy Performance of Buildings (recast)," *EUR-Lex* (Official Journal of the European Union), 2010, https://eur-lex.europa.eu/legal-content/en/TXT/?uri=celex%3A32010L0031.

80. European Commission website, https://ec.europa.eu/energy/en/topics/energy-efficiency /energy-performance-of-buildings, accessed December 18, 2018.

81. Jollands et al., 2010.

82. Weiss, Dunkelberg, and Vogelpohl, 2012.

83. Jollands et al., 2010.

84. S. Bin and S. Nadel, "How Does China Achieve a 95 Percent Compliance Rate for Building Energy Codes? A Discussion About China's Inspection System and Compliance Rates," *2012 ACEEE Summer Study on Energy-Efficiency in Buildings*, 2012, https://aceee.org/files/proceedings/2012/data/papers/0193-000261.pdf.

85. Xiaojing Sun et al., "Mandating Better Buildings: A Global Review of Building Codes and Prospects for Improvement in the United States," *Wiley Interdisciplinary Reviews (WIREs): Energy and Environment* 5, no. 2 (2016): 188–215, doi:10.1002/wene.168.

86. Jollands et al., 2010.

87. Bemelmans-Videc, Rist, and Vedung, 2017.

88. Renewable Energy Policy Network for the 21st Century (REN21), *Renewables 2018 Global Status Report*, http://www.ren21.net/gsr-2018/.

89. Oliver Frank (German Energy Agency, DENA), "Renewable Energies in Germany" (presentation to the Energy Study Tour, Berlin, December 11, 2014).

90. S. V. Valentine, *Wind Power: Politics and Policy* (Oxford: Oxford University Press, 2015).

91. For more on this development, see http://www.nortonrosefulbright.com/knowledge /publications/147727/german-renewable-energy-act-2017-eeg-2017-what-you-should -know.

92. International Energy Agency database, https://www.iea.org/policiesandmeasures /pams/india/name-140460-en.php, accessed January 8, 2018.

93. Natural Resources Canada website, http://www.nrcan.gc.ca/energy/fuel-prices/18885.

94. https://www.ato.gov.au/business/excise-and-excise-equivalent-goods/fuel-excise /excise-rates-for-fuel/.

95. https://www.gov.uk/tax-on-shopping/fuel-duty.

96. International Energy Agency database, https://www.iea.org/policiesandmeasures /pams/israel/name-165079-en.php, accessed January 8, 2018.

97. International Energy Agency database, https://www.iea.org/policiesandmeasures /pams/australia/name-162382-en.php, accessed January 8, 2018.

98. International Energy Agency database, https://www.iea.org/policiesandmeasures /pams/india/name-159041-en.php, accessed December 18, 2018.

99. International Energy Agency database, https://www.iea.org/policiesandmeasures /pams/mexico/name-24376-en.php, accessed December 18, 2018.

100. Singapore National Environmental Agency website, http://www.nea.gov.sg/energy -waste/energy-efficiency/industry-sector, accessed January 8, 2018.

101. More on the E2F program at https://www.e2singapore.gov.sg/incentives/energy -efficiency-fund.

102. M. Howlett, M. Ramesh, and A. Perl, *Studying Public Policy: Policy Cycles and Policy Subsystems*, 3rd ed. (Oxford: Oxford University Press, 2009).

103. Valentine, 2011.

104. Valentine, 2014.

105. Louisa Tang, "Studying Behavior 'Can Lead to Better Results for Schemes, Policies,'" *Today* (Singapore), June 26, 2015, accessed January 10, 2018, http://www.todayonline.com /singapore/online-space-offers-great-scope-shape-citizens-behaviour-civil-service-head.

106. More on this scheme at https://travelsmartrewards.lta.gov.sg/, accessed May 4, 2019.

107. Toby Park, *Evaluating the Nest Learning Thermostat*, The Behavioral Insights Team, November 30, 2017, accessed January 10, 2018, http://www.behaviouralinsights.co .uk/publications/evaluating-the-nest-learning-thermostat/.

108. Valentine, 2015.

109. More on these initiatives at http://ec.europa.eu/research/industrial_technologies/ppp -in-research_en.html, accessed January 10, 2018.

110. K. Tanaka, "Review of Policies and Measures for Energy Efficiency in Industry Sector," *Energy Policy* 39 (2011): 6532–6550.

111. Visit its website at https://www.efficiencyvermont.com/services.

112. Jason Erwin et al., "Final Report: 'Case Study on Evaluation of Energy Efficiency Information Centers and One-Stop Shops" (prepared for WEC/ADEME preparations for WEC Congress, Istanbul, October 2016).

113. For more on this subject, see http://www.ifc.org/wps/wcm/connect/41f4e300407f54ed 851595cdd0ee9c33/SectorSheets_Renewables.pdf?MOD=AJPERES.

114. David Hatch, "Japan 'Smart' Cities Rely on Public-Private Partnerships," *CitiSignals*, September 30, 2016, accessed May 4, 2019, http://archive.citiscope.org/citisignals /2016/japan-smart-cities-rely-public-private-partnerships.

115. See the Perform Achieve Trade in India: http://iepd.iipnetwork.org/policy/perform -achieve-trade-scheme-pat-scheme.

116. Climate and Development Knowledge Network (CDKN), *Creating Market Support for Energy Efficiency: India's Perform, Achieve and Trade Scheme*, 2013, http://cdkn .org/wp-content/uploads/2013/01/India-PAT_InsideStory.pdf.

7. Consumers: From a Source of Problems to Agents of Change

1. Aaron M. McCright et al., "Ideology, Capitalism, and Climate: Explaining Public Views About Climate Change in the United States," *Energy Research and Social Science* 21 (2016): 180–189; Steven R. Brechin and Medani Bhandari, "Perceptions of Climate Change Worldwide," *Wiley Interdisciplinary Reviews (WIREs): Climate Change* 2 (2011): 871–885.

2. Paul C. Stern et al., "Opportunities and Insights for Reducing Fossil Fuel Consumption by Households and Organizations," *Nature Energy* 1 (2016): 16043, doi:10.1038/nenergy.2016.43.

3. Benjamin K. Sovacool, Marilyn A. Brown, and Scott Valentine, *Fact and Fiction in Global Energy Policy* (Baltimore, MD: Johns Hopkins University Press, 2016); P. Hawken, A. Lovins, and L. H. Lovins, *Natural Capitalism: Creating the Next Industrial Revolution* (New York: Little, Brown, 1999).

4. R. Kok, René M. J. Benders, and Henri C. Moll, "Measuring the Environmental Load of Household Consumption Using Some Methods Based on Input–Output Energy Analysis: A Comparison of Methods and a Discussion of Results," *Energy Policy* 34 (2006): 2744–2761.

5. United Nations Environment Programme (UNEP), *Kick the Habit: A UN Guide to Climate Neutrality* (Paris: UNEP, 2008).

6. P. Kivimaa and M. Martiskainen, "Innovation, Low Energy Buildings and Intermediaries in Europe: Systematic Case Study Review," *Energy Efficiency* 11, no. 1 (2018): 31–51.

7. J. Schot, L. Kanger, and G. Verbong, "The Roles of Users in Shaping Transitions to New Energy Systems," *Nature Energy* 16054 (2016).

8. Y. Parag and B. K. Sovacool, "Electricity Market Design for the Prosumer Era," *Nature Energy* 16032 (2016): 1–6.

9. Marilyn A. Brown, Shan Zhou, and Majid Ahmadi, "Smart Grid Governance: An International Review of Evolving Policy Issues and Innovations," *Wiley Interdisciplinary Reviews (WIREs): Energy and Environment* 7, no. 5 (2018): e290.

10. E. M. Rogers, *Diffusion of Innovations*, 5th ed. (New York: Simon and Schuster, 2003).

11. Rogers, 2003, 11.

12. Kimberly S. Wolske, Paul C. Stern, and Thomas Dietz, "Explaining Interest in Adopting Residential Solar Photovoltaic Systems in the United States: Toward an Integration of Behavioral Theories," *Energy Research and Social Science* 25 (2017): 134–151.

13. A. Faiers and C. Neame, "Consumer Attitudes Towards Domestic Solar Power Systems," *Energy Policy* 34 (2006): 1797–1806, http://dx.doi.org/10.1016/j.enpol.2005.01.001.

14. B. Bollinger and K. Gillingham, "Peer Effects in the Diffusion of Solar Photovoltaic Panels," *Marketing Science* 31, no. 6 (2012): 900–912, http://dx.doi.org/10.1287/mksc.1120.0727.

15. V. Rai and S.A. Robinson, "Effective Information Channels for Reducing Costs of Environmentally-Friendly Technologies: Evidence from Residential PV Markets," *Environmental Research Letters* 8 (2013): 014044.

16. Erik van der Vleuten, "Understanding Network Societies: Two Decades of Large Technical Systems Studies," in *Networking Europe: Transnational Infrastructures and the Shaping of Europe, 1850–2000*, ed. E. van der Vleuten and A. Kaijser (Sagamore Beach, MA: Science History Publications, 2006), 279–314.

17. Arnulf Grubler, Charlie Wilson, and Gregory Nemet, "Apples, Oranges, and Consistent Comparisons of the Temporal Dynamics of Energy Transitions," *Energy Research and Social Science* 22 (2016): 18–25.

18. Thomas P. Hughes, *Networks of Power: Electrification in Western Society, 1880–1930* (Cambridge, MA: MIT Press, 1983), 15.

19. Bernward Joerges, "Large Technical Systems: Concepts and Issues," in *The Development of Large Technical Systems*, ed. Renate Mayntz and Thomas P. Hughes (Boulder, CO: Westview Press, 1988), 9–36; Marilyn A. Brown et al., Carbon Lock-In: Barriers to the Deployment of Climate Change Mitigation Technologies (Oak Ridge, TN: Oak Ridge National Laboratory, 2008), https://www.acs.org/content/dam/acsorg/policy/acsonthehill/briefings/solarenergy/report-carbon-lock-in.pdf; Marilyn A. Brown and Sharon (Jess) Chandler, "Governing Confusion: How Statutes, Fiscal Policy, and Regulations Impede Clean Energy Technologies," *Stanford Law and Policy Review* 19, no. 3 (2008): 472–509.

20. Arie Rip, "Introduction of New Technology: Making Use of Recent Insights from Sociology and Economics of Technology," *Technology Analysis and Strategic Management* 7, no. 4 (1995): 417–443.

21. S. M. Macey and M. A. Brown, "Residential Energy Conservation Through Repetitive Household Behaviors," *Environment and Behavior* 15 (1983): 123–141.

22. E. Shove, M. Pantzar, and M. Watson, *The Dynamics of Social Practice: Everyday Life and How It Changes* (London: Sage, 2012).

23. I. Ropke, "Theories of Practice: New Inspiration for Ecological Economic Studies of Consumption," *Ecological Economics* 68, no. 10 (2009): 2490–2497; A. Reckwitz, "Toward a Theory of Social Practices: A Development in Culturalist Theorizing," *European Journal of Social Theory* 5, no. 2 (2002): 243–263; M. Watson, "How Theories of Practice Can Inform Transition to a Decarbonised Transport System," *Journal of Transport Geography* 24 (2012): 488–496.

24. B. K. Sovacool and D. J. Hess, "Ordering Theories: Typologies and Conceptual Frameworks for Sociotechnical Change," *Social Studies of Science* 47, no. 5 (2017): 703–750.

25. T. Jackson, *Motivating Sustainable Consumption: A Review of Evidence on Consumer Behaviour and Behavioural Change* (Surrey, UK: Sustainable Development Research Network, 2005), http://www.sustainablelifestyles.ac.uk/sites/default/files/motivating_sc_final.pdf.

26. T. Dietz, "Environmental Value," in *Handbook of Value: Perspectives from Economics, Neuroscience, Philosophy, Psychology and Sociology*, ed. T. Brosch and D. Sander (Oxford: Oxford University Press, 2015), 329–349.

27. T. Dietz and P. C. Stern, "Toward a Theory of Choice: Socially Embedded Preference Construction," *Journal of Socio-Economics* 24, no. 2 (1995): 261–279.

28. Macey and Brown, 1983.

29. D. Jamieson, "Ethics, Public Policy, and Global Warming," in *Climate Ethics: Essential Readings*, ed. S. M. Gardiner et al. (Oxford: Oxford University Press, 2010): 147.

30. B. K. Sovacool and S. V. Valentine, *The National Politics of Nuclear Power: Economics, Security, and Governance* (New York: Routledge, 2012).

31. L. Steg et al., "The Significance of Hedonic Values for Environmentally Relevant Attitudes, Preferences, and Actions," *Environment and Behavior* 46, no. 2 (2014):

163–192; P. C. Stern, T. Dietz, and G. A. Guagnano, "A Brief Inventory of Values," *Educational and Psychological Measurement* 58 (1998): 884–1001.

32. Marilyn A. Brown and Benjamin K. Sovacool, "Theorizing the Behavioral Dimension of Energy Consumption: Energy Efficiency and the Value-Action Gap," in *The Oxford Handbook of Energy and Society*, ed. Debra J. Davidson and Matthias Gross (Oxford: Oxford University Press, 2018).

33. I. Ajzen and M. Fishbein, *Understanding Attitudes and Predicting Social Behavior* (Englewood Cliffs, NJ: Prentice Hall, 1980); Macey and Brown, 1983.

34. P. A. Sabatier, "An Advocacy Coalition Framework of Policy Change and the Role of Policy-Oriented Learning Therein," *Policy Sciences* 21, no. 2 (1988): 129–168.

35. Reza Kowsari and Hisham Zerriffi, "Three Dimensional Energy Profile: A Conceptual Framework for Assessing Household Energy Use," *Energy Policy* 39, no. 12 (2011): 7505–7517.

36. Walt Patterson, *Keeping the Lights On: Towards Sustainable Electricity* (London: Earthscan, 2007).

37. Paul C. Stern and Elliot Aronson, *Energy Use: The Human Dimension* (New York: Freeman, 1984).

38. Martin J. Pasqualetti, "Morality, Space, and the Power of Wind-Energy Landscapes," *Geographical Review* 90, no. 3 (2000): 384 and 386.

39. Stern and Aronson, 1984.

40. James C. Williams, "Strictly Business: Notes on Deregulating Electricity," *Technology and Culture* 42 (2001): 626–630.

41. Paul N. Edwards, "Infrastructure and Modernity: Force, Time, and Social Organization in the History of Sociotechnical Systems," in *Modernity and Technology*, ed. Thomas J. Misa, Philip Brey, and Andrew Feenberg (Cambridge, MA: MIT Press, 2003), 185–186.

42. Christopher Groves et al., "The Grit in the Oyster: Using Energy Biographies to Question Socio-Technical Imaginaries of 'Smartness,'" *Journal of Responsible Innovation* 3, no. 1 (2016): 4–25.

43. Christina Liddell, "Human Factors in Energy Efficient Housing: Insights from a Northern Ireland Pocket Neighbourhood," *Energy Research and Social Science* 10 (2015): 19–25.

44. Quoted in Benjamin K. Sovacool, *The Dirty Energy Dilemma: What's Blocking Clean Power in the United States* (Westport, CT: Praeger, 2008).

45. National Environmental Education and Training Foundation (NEETF) and Roper ASW, *Americans' Low "Energy IQ": A Risk to Our Energy Future: Why America Needs a Refresher Course on Energy*, The Tenth Annual National Report Card: Energy Knowledge, Attitudes, and Behavior (Washington, DC: NEETF, August 2002).

46. Kentucky Environmental Education Council (KEEC), *The 2004 Survey of Kentuckians' Environmental Knowledge, Attitudes and Behaviors* (Frankfort, KY: KEEC, January 2005).

47. Suzanne Crofts Shelton, *The Consumer Pulse Survey on Energy Conservation* (Knoxville, TN: Shelton Group, 2006).

48. Glenn Hess, "Bush Promotes Alternative Fuel," *Chemical and Engineering News*, March 6, 2006, 50–58.

49. B. K. Sovacool, "Differing Cultures of Energy Security: An International Comparison of Public Perceptions," *Renewable and Sustainable Energy Reviews* 55 (2016): 811–822.

50. F. Dianshu, B. K. Sovacool, and M. V. Khuong, "The Barriers to Energy Efficiency in China: Assessing Household Electricity Savings and Consumer Behavior in Liaoning Province," *Energy Policy* 38, no. 2) (2010): 1202–1209.

51. B. K. Sovacool and I. M. Drupady, *Energy Access, Poverty, and Development: The Governance of Small-Scale Renewable Energy in Developing Asia* (New York: Ashgate, 2012).

52. Sovacool and Drupady, 2012.

53. Sovacool and Drupady, 2012.

54. Sovacool and Drupady, 2012.

55. See Jack N. Barkenbus, "Eco-Driving: An Overlooked Climate Change Initiative," *Energy Policy* 38, no. 2 (2010): 762–769; the Vanderbilt team's research is in Amanda R. Carrico et al., "Costly Myths: An Analysis of Idling Beliefs and Behavior in Personal Motor Vehicles," *Energy Policy* 37, no. 8 (2009): 2881–2888.

56. B. K. Sovacool, "Solving the Oil Independence Problem: Is It Possible?" *Energy Policy* 35, no. 11 (2007): 5505–5514.

57. B. K. Sovacool, "The Cultural Barriers to Renewable Energy in the United States," *Technology in Society* 31, no. 4 (2009): 365–373.

58. Martin V. Melosi, *Coping with Abundance: Energy and Environment in Industrial America* (New York: Alfred A. Knopf, 1985), 8–10.

59. Langdon Winner, "Energy Regimes and the Ideology of Efficiency," in *Energy and Transport: Historical Perspectives on Policy Issues*, ed. George H. Daniels and Mark H. Rose (London: Sage, 1982), 261–277.

60. Dorothy K. Newman and Don Day, *The American Energy Consumer* (Cambridge, MA: Ballinger, 1975).

61. Bonnie Mass Morrison and Peter Gladhart, "Energy and Families: The Crisis and Response," *Journal of Home Economics* 68, no. 1 (1976): 15–18.

62. Sovacool and Drupady, 2012.

63. See Suzanne C. Thompson, "Will It Hurt Less If I Control It? A Complex Answer to a Simple Question," *Psychological Bulletin* 90, no. 1 (1981): 89–101; Sharon S. Brehm and Jack W. Brehm, *Psychological Reactance: A Theory of Freedom and Control* (New York: Academic Press, 1981); Michael B. Mazis, "Antipollution Measures and Psychological Reactance Theory: A Field Experiment," *Journal of Personality and Social Psychology* 31, no. 4 (1975): 654–660; Lawrence J. Becker, "Joint Effect of Feedback and Goal Setting on Performance: A Field Study of Residential Energy Conservation," *Journal of Applied Psychology* 63, no. 4 (1978): 428–433.

64. Paul C. Stern and Elliot Aronson, *Energy Use: The Human Dimension* (New York: Freeman, 1984).

65. L. J. Becker, C. Seligman, and J. M. Darley, *Psychological Strategies to Reduce Energy Consumption* (Princeton, NJ: Center for Energy and Environmental Studies, 1979).

66. Ellen J. Langer and Judith Rodin, "The Effects of Choice and Enhanced Personal Responsibility for the Aged: A Field Experiment in an Institutional Setting," *Journal of Personality and Social Psychology* 34, no. 2 (1978): 191.

67. Jason Chilvers and Noel Longhurst, "Participation in Transition(s): Reconceiving Public Engagements in Energy Transitions as Co-Produced, Emergent and Diverse," *Journal of Environmental Policy and Planning* 18, no. 5 (2016): 585–607.

68. Nina Kahma and Kaisa Matschoss, "The Rejection of Innovations? Rethinking Technology Diffusion and the Non-Use of Smart Energy Services in Finland," *Energy Research and Social Science* 34 (2017): 27–36.

69. K. Buchanan et al., "The British Public's Perception of the UK Smart Metering Initiative: Threats and Opportunities," *Energy Policy* 91 (2016): 87–97; B. K. Sovacool et al., "Vulnerability and Resistance in the United Kingdom's Smart Meter Transition," *Energy Policy* 109 (2017): 767–781.

70. Nazmiye Balta-Ozkan, Oscar Amerighi, and Benjamin Boteler, "A Comparison of Consumer Perceptions Towards Smart Homes in the UK, Germany and Italy: Reflections for Policy and Future Research," *Technology Analysis and Strategic Management* 26, no. 10 (2014): 1176–1195.

71. Balta-Ozkan, Amerighi, and Boteler, 2014.

72. J. Savirimuthu, "Smart Meters and the Information Panopticon: Beyond the Rhetoric of Compliance," *International Review of Law, Computers and Technology* 27, nos.1–2 (2013): 161–186.

73. David J. Hess, "Smart Meters and Public Acceptance: Comparative Analysis and Governance Implications," *Health, Risk and Society* 16, no. 3 (2014): 243–258.

74. Parag and Sovacool, 2016.

75. *Home Automation (Luxury, Mainstream, DIY [Do It Yourself] and Managed) Market by Networking Technology (Wired, Power-line, Computing Network and Wireless) for Lighting, Safety and Security, HVAC, Entertainment and Other (Robotics and Heath Care) Applications—Global Industry Perspective, Comprehensive Analysis, and Forecast, 2014–2020* (Sarasota, FL: Zion Market Research, 2015), https://www.market-researchstore.com/report/home-automation-market-z38751

76. Parag and Sovacool, 2016.

77. https://vandebron.nl/about.

78. R. Hiteva and B. K. Sovacool, "Harnessing Social Innovation for Energy Justice: A Business Model Perspective," *Energy Policy* 107 (2017): 631–639.

79. *Green Mountain Power Customers Help Transform Vermont's Energy Future* (Colchester, VT: Green Mountain Power, 2017), https://www.greenmountainpower.com/press/green-mountain-power-customers-help-transform-vermonts-energy-future/.

80. Tesla official website, https://www.tesla.com/green-mountain-power.

81. Diane Cardwell, "Utility Helps Wean Vermonters from the Electric Grid," *New York Times*, July 29, 2017, https://www.nytimes.com/2017/07/29/business/energy-environment/vermont-green-mountain-power-grid.html.

82. Green Mountain Power, "In-Home Level 2 EV Charger," https://greenmountain power.com/product/home-level-2-ev-charger/.

83. *Green Mountain Power and SunCommon Organize First Community Solar Array for Low Income Vermonters* (Colchester, CT: Green Mountain Power, 2017), https://www.greenmountainpower.com/press/green-mountain-power-suncommon-organize-first-community-solar-array-low-income-vermonters/.

84. B. K. Sovacool, "The Importance of Open and Closed Styles of Energy Research," *Social Studies of Science* 40, no. 6 (2010): 903–930; M. A. Brown and B. K. Sovacool, "Brazil's Proalcohol Program and Promotion of Flex-Fuel Vehicles," in *Climate Change and Global Energy Security: Technology and Policy Options* (Cambridge, MA: MIT Press, 2011), 260–274.

85. Juan Forero, "Brazil's Ethanol Sector, Once Thriving, Is Being Buffeted by Forces Both Man-Made, Natural," *Washington Post*, January 1, 2014, https://www.washingtonpost.com/world/brazils-ethanol-sector-once-thriving-is-being-buffeted-by-forces-both-man-made-natural/2014/01/01/9587b416-56d7-11e3-bdbf-097ab2a3dc2b_story.html?utm_term=.ed5eb84688e5.

8. Minimizing Governance Barriers

1. Benjamin K. Sovacool and Charmaine Watts, "Going Completely Renewable: Is It Possible (Let Alone Desirable)?" *Electricity Journal* 22, no. 4 (2009): 95–111.

2. Sovacool and Watts, 2009.

3. Benjamin K. Sovacool, *The Dirty Energy Dilemma* (Westport, CT: Praeger, 2008), 113–121; Mark Z. Jacobson, "Review of Solutions to Global Warming, Air Pollution, and Energy Security," *Energy and Environmental Science* 2 (2009): 148–173.

4. Paraphrased from a presentation made by Mark Diesendorf during a workshop at City University of Hong Kong, 2013.

5. Windmonitor Frauhofer website, accessed December 12, 2017, http://windmonitor.iwes.fraunhofer.de/windmonitor_en/1_wind-im-strommix/1_energiewende-in-deutschland/6_Ausbaustand_der_Bundeslaender/.

6. Arthur Neslen, "Wind Power Generates 140 Percent of Denmark's Electricity Demand," *The Guardian*, July 10, 2015, accessed December 12, 2017, https://www.theguardian.com/environment/2015/jul/10/denmark-wind-windfarm-power-exceed-electricity-demand.

7. Peter Walker, "Denmark Runs Entirely on Wind Energy for a Day," *The Independent*, March 2, 2017, accessed December 12, 2017, http://www.independent.co.uk/news/world/europe/denmark-ran-entirely-on-wind-energy-for-a-day-a7607991.html.

8. L. W. Ho, "Wind Energy in Malaysia: Past, Present and Future," *Renewable and Sustainable Energy Reviews* 53 (2016): 279–295; B. K. Sovacool and I. M. Drupady, "Examining the Small Renewable Energy Power (SREP) Program in Malaysia," *Energy Policy* 39, no. 11 (2011): 7244–7256.

9. Elinor Ostrom, *A Polycentric Approach for Coping with Climate Change*, Policy Research working paper no. WPS 5095 (Washington, DC: World Bank, 2009).

10. A. E. Florini and B. K. Sovacool. "Bridging the Gaps in Global Energy Governance," *Global Governance* 17, no. 1 (2011): 57–74; A. E. Florini and B. K. Sovacool, "Who Governs Energy? The Challenges Facing Global Energy Governance," *Energy Policy* 37, no. 12 (2009): 5239–5248.

11. Ann Florini, *The Coming Democracy: New Rules for Running a New World* (Washington, DC: Brookings Institution, 2005).

12. The biggest divestment to date has been made by Norway's sovereign wealth fund. For further information, see https://www.nytimes.com/2017/11/16/business/energy-environment/norway-fund-oil.html.

13. Paul A. Samuelson, "The Pure Theory of Public Expenditure," *Review of Economics and Statistics* 36 (1954): 387–389.

14. Florini and Sovacool, 2009.

15. B. K. Sovacool and H. Saunders, "Competing Policy Packages and the Complexity of Energy Security," *Energy* 67 (2014): 641–651.

16. Steven J. Davis and Ken Caldeira, "Consumption-Based Accounting of CO_2 Emissions," *PNAS* 107, no. 12 (2010): 1–6.

17. Steven J. Davis, Glen P. Peters, and Ken Caldeira, "The Supply Chain of CO_2 Emissions," *PNAS* 108, no. 45 (2011): 18554–18559.

18. United States Environmental Protection Agency, *Mercury Study Report to Congress: Volume I: Executive Summary*, EPA-452/R-97-003, December 1997.

19. Michael Hopkin, "Acid Rain Still Hurting Canada," *Nature*, August 20, 2005.

20. Erik Behrens et al., "Model Simulations on the Long-Term Dispersal of 137Cs Released into the Pacific Ocean off Fukushima," *Environmental Research Letters* 7, no. 3 (2012): 034004.

21. Steven Lee Myers, "Lament for a Once-Lovely Waterway," *New York Times*, June 12, 2010; Steven Lee Myers, "Vital River Is Withering, and Iraq Has No Answer," *New York Times*, June 12, 2010.

22. J. Jackson Ewing and Elizabeth McRae, "Transboundary Haze in Southeast Asia: Challenges and Pathways Forward," *NTS Alert*, October 2012.

23. For more on this subject, see the WTO site, https://www.wto.org/english/thewto_e/whatis_e/tif_e/fact2_e.htm.

24. Benjamin K. Sovacool, "Placing a Glove on the Invisible Hand: How Intellectual Property Rights May Impede Innovation in Energy Research and Development (R&D)," *Albany Law Journal of Science and Technology* 18, no. 2 (2008): 381–440.

25. Marilyn A. Brown et al., *Carbon Lock-In: Barriers to the Deployment of Climate Change Mitigation Technologies* (Oak Ridge, TN: Oak Ridge National Laboratory, 2007).

26. David Coady et al., *How Large Are Global Energy Subsidies?* IMF Working Paper WP/15/105, May 2015, http://www.imf.org/external/pubs/ft/wp/2015/wp15105.pdf.

27. Coady et al., 2015.

28. International Energy Agency (IEA), *World Energy Outlook 2008* (Paris: OECD/IEA, 2008).

29. IEA, 2008.

30. Ronald Steenblik, *Biofuels—At What Cost? Government Support for Ethanol and Biodiesel in Selected OECD Countries* (Geneva: International Institute for Sustainable Development, September 2007).

31. B. K. Sovacool, "Reviewing, Reforming, and Rethinking Global Energy Subsidies: Towards a Political Economy Research Agenda," *Ecological Economics* 135 (2017): 150–163.

32. Steenblik, 2007.

33. Doug Koplow and John Dernbach, "Federal Fossil Fuel Subsidies and Greenhouse Gas Emissions: A Case Study of Increasing Transparency for Fiscal Policy," *Annual Review of Energy and Environment* 26 (2001): 361–389.

34. Douglas Koplow, "Global Energy Subsidies: Scale, Opportunity Costs, and Barriers to Reform," in *Energy Poverty: Global Challenges and Local Solutions*, ed. Antoine Halff, Benjamin K. Sovacool, and Jon Rozhon (Oxford: Oxford University Press, 2014).

35. Sadeq Z. Bigdeli, "Will the Friends of Climate Emerge in the WTO? The Prospects of Applying the Fisheries Subsidy Model to Energy Subsidies," *Carbon and Climate Law Review* 2, no. 1 (2008): 78–88.

36. Andrew Simms, *The Price of Power: Poverty, Climate Change, the Coming Energy Crisis, and the Renewable Revolution* (London: New Economics Foundation, 2004).

37. Cees van Beers and Andre de Moor, *Public Subsidies and Policy Failures: How Subsidies Distort the Natural Environment, Equity, and Trade, and How to Reform Them* (Cheltenham: Edward Elgar, 2001).

38. Masami Kojima, *Government Responses to Oil Price Volatility*, Extractive Industries for Development Series No. 10 (Washington, DC: World Bank, July 2009).

39. Brian Head, "Wicked Problems in Public Policy," *Public Policy* 3, no. 2 (2008): 101–118; Kelly Levin et al., "Overcoming the Tragedy of Super Wicked Problems: Constraining Our Future Selves to Ameliorate Global Climate Change," *Policy Sciences* 45, no. 2 (2012): 123–152.

40. World Wide Fund for Nature (WWF), *Certification and Roundtables: Do They Work? WWF Review of Multi-Stakeholder Sustainability Initiatives*, September 2010, https://d2ouvy59p0dg6k.cloudfront.net/downloads/wwf_msireview_sept_2010_lowres.pdf.

41. Dan Klooster, "Standardizing Sustainable Development? The Forest Stewardship Council's Plantation Policy Review Process as Neoliberal Environmental Governance," *Geoforum* 41 (2010): 117–129; Klaus Dingwerth, "North-South Parity in Global Governance: The Affirmative Procedures of the Forest Stewardship Council," *Global Governance* 14 (2008): 53–71; Stephen Bass et al., *Certification's Impacts on Forests, Stakeholders and Supply Chains*, Instruments for Sustainable Private Sector Forestry Series (London: International Institute for Environment and Development, 2001); Peter Leigh Taylor, "In the Market but Not of It: Fair Trade Coffee and Forest Stewardship Council Certification as Market-Based Social Change," *World Development* 33, no. 1 (2005): 129–147; J. Chen, J. L. Innes, and A. Tikina, "Private Cost-Benefits of Voluntary Forest Product Certification," *International Forestry Review* 12, no. 1 (2010):

1–12; Ralph Espach, "When Is Sustainable Forestry Sustainable? The Forest Stewardship Council in Argentina and Brazil," *Global Environmental Politics* 6, no. 2 (2006): 55–84.

42. William F. Laurance et al., "Improving the Performance of the Roundtable on Sustainable Palm Oil for Nature Conservation," *Conservation Biology* 24, no. 2 (2010): 377–381; Greetje Schouten and Pieter Glasbergen, "Creating Legitimacy in Global Private Governance: The Case of the Roundtable on Sustainable Palm Oil," *Ecological Economics* 70 (2011): 1891–1899.

43. Elizabeth Fortin and Ben Richardson, "Certification Schemes and the Governance of Land: Enforcing Standards or Enabling Scrutiny?" *Globalizations* 10, no. 1 (2013): 141–159.

44. Patrick Anderson, *Free, Prior, and Informed Consent in REDD+: Principles and Approaches for Policy and Project Development* (Bangkok: RECOFTC and GIZ, 2011); Ann E. Prouty, "The Clean Development Mechanisms and Its Implications for Climate Justice," *Columbia Journal of Environmental Law* 34, no. 2 (2009): 513–540; Benjamin K. Sovacool and Bjorn-Ola Linnér, *The Political Economy of Climate Change Adaptation* (Basingstoke, UK: Palgrave Macmillan, 2015).

45. Benjamin K. Sovacool and Nathan Andrews, "Does Transparency Matter? Evaluating the Governance Impacts of the Extractive Industries Transparency Initiative (EITI) in Azerbaijan and Liberia," *Resources Policy* 45 (2015): 183–192.

46. Benjamin K. Sovacool et al., "Energy Governance, Transnational Rules, and the Resource Curse: Exploring the Effectiveness of the Extractive Industries Transparency Initiative (EITI)," *World Development* 86 (2016): 179–192.

47. Kenneth Burke, "How Many Texts Do People Send Every Day (2018)?" *Text Request* (blog), May 18, 2016, accessed December 18, 2018, https://www.textrequest.com/blog/how-many-texts-people-send-per-day/.

48. By the mid-1990s, over 50 million people were using the Internet, according to http://en.wikimedia.org/wiki/File:Internet_users_per_100_inhabitants_ITU.svg.

49. http://www.theguardian.com/news/datablog/2014/feb/04/facebook-in-numbers-statistics.

50. "How Many Google Searches per Day on Average in 2018?" *Ardor SEO* (blog), accessed December 18, 2018, https://ardorseo.com/blog/how-many-google-searches-per-day-2018/.

51. "Internet of Things (IoT) Connected Devices Installed Base Worldwide from 2015 to 2025 (in Billions)," *Statista*, accessed December 18, 2018, https://www.statista.com/statistics/471264/iot-number-of-connected-devices-worldwide/.

52. "Number of Monthly Active Twitter Users Worldwide from First Quarter 2010 to Fourth Quarter 2018 (in Millions)," *Statista*, accessed December 18, 2018, https://www.statista.com/statistics/282087/number-of-monthly-active-twitter-users/.

53. Carl Hedde, *2013 Natural Catastrophe Year in Review* (Princeton, NJ: Munich Reinsurance America, 2014).

54. IPCC, 2012: *Managing the Risks of Extreme Events and Disasters to Advance Climate Change Adaptation. A Special Report of Working Groups I and II of the*

Intergovernmental Panel on Climate Change, ed. C. B. Field et al. (Cambridge: Cambridge University Press, 2012).

55. Krister P. Andersson and Elinor Ostrom, "Analyzing Decentralized Resource Regimes from a Polycentric Perspective," *Policy Sciences* 41 (2008): 71–93.

56. E. Ostrom, "Polycentric Systems for Coping with Collective Action and Global Environmental Change," *Global Environmental Change* 20, no. 4 (2010): 550–557.

57. B. K. Sovacool and M. A. Brown, "Scaling the Response to Climate Change," *Policy and Society* 27, no. 4 (2009): 317–328.

58. David Roberts, "For the First Time, a Major US Utility Has Committed to 100 Percent Clean Energy," *Vox*, December 14, 2018, accessed December 18, 2018, https://www.vox.com/energy-and-environment/2018/12/5/18126920/xcel-energy -100-percent-clean-carbon-free.

59. B. K. Sovacool, "An International Comparison of Four Polycentric Approaches to Climate and Energy Governance," *Energy Policy* 39, no. 6 (2011): 3832–3844.

60. S. V. Valentine, *Wind Power: Politics and Policy* (Oxford: Oxford University Press, 2014).

61. M. A. Brown and B. K. Sovacool, *Climate Change and Global Energy Security: Technology and Policy Options* (Cambridge, MA: MIT Press, 2011).

62. S. V. Valentine, "Gradualist Best Practice in Wind Power Policy," *Energy for Sustainable Development* 22 (2014): 74–84.

63. Brown and Sovacool, 2011.

64. Brown and Sovacool, 2011.

65. Centre for Public Impact website, accessed December 18, 2018, https://www.centre forpublicimpact.org/case-study/solar-home-systems-bangladesh/.

66. Statistics drawn from https://www.statista.com/statistics/278566/urban-and-rural -population-of-china/.

67. Brown and Sovacool, 2011.

68. Brown and Sovacool, 2011.

69. Brown and Sovacool, 2011.

70. Brown and Sovacool, 2011.

71. Georgina Santos, Wai Wing Li, and Winston T. H. Koh, "Transport Policies in Singapore," in *Road Pricing: Theory and Evidence*, ed. Georgina Santos (Oxford: Elsevier, 2004), 209–235.

72. Leo Tan Wee Hin and R. Subramaniam, "Congestion Control of Heavy Vehicles Using Electronic Road Pricing: The Singapore Experience," *International Journal of Heavy Vehicle Systems* 13, nos. 1–2 (2006): 37–55.

73. Paul A. Barter, "Singapore's Urban Transport: Sustainability by Design or Necessity?" in *Spatial Planning for a Sustainable Singapore*, ed. Tai-Chee Wong, Belinda Yuen, and Charles Goldblum (New York: Springer, 2008), 95–112; Santos et al., 2004.

74. Brown and Sovacool, 2011.

75. Brown and Sovacool, 2011.

76. Brown and Sovacool, 2011.

77. California Energy Commission, *2012 Accomplishments*, https://www.energy.ca.gov /commission/accomplishments/2012_energycommisson_accomplishments.pdf.

78. Patrick Saxton, *CALGreen, Title 24 and the Net Zero Energy Standard* (Sacramento, CA: California Energy Commission, 2012), http://www.green-technology.org /gcsummit/images/CALGreen-Title-24.pdf.

79. *California's 2016 Residential Building Energy Efficiency Standards* (Sacramento, CA: California Energy Commission, 2016), http://www.energy.ca.gov/title24/2016standards /rulemaking/documents/2016_Building_Energy_Efficiency_Standards_infographic .pdf.

80. http://www.csmonitor.com/USA/2010/0115/California-adopts-firststatewide -green-building-code.

81. American Council for an Energy-Efficient Economy (ACEEE), *California*, 2018, https://database.aceee.org/state/california.

82. American Council for an Energy-Efficient Economy (ACEEE), *Los Angeles, CA*, 2018, https://database.aceee.org/city/los-angeles-ca.

83. "Los Angeles Tops List of Cities with Most Energy Efficient Buildings," *CBS Los Angeles*, April 10, 2014.

84. American Council for an Energy-Efficient Economy (ACEEE), *San Francisco, CA*, 2018, https://database.aceee.org/city/san-francisco-ca.

85. More on the City of Berkeley's programs at *Energy Efficiency* (Berkeley, CA: Office of Energy and Sustainable Development, n.d.). https://www.cityofberkeley.info /EnergyEfficiencyPrograms/.

86. *Berkeley Climate Action Plan: Tracking Our Progress: Building Energy Use—Solar PV* (Berkeley, CA: Office of Energy and Sustainable Development, April 2014). Accessed on May 4, 2019 at https://www.cityofberkeley.info/climate/.

87. City of Oakland, California, "Energy Programs Serving Oakland," Facilities and Environment, 2014.

88. American Council for an Energy-Efficient Economy (ACEEE), *Oakland, CA*, Community-Wide Initiatives, State and Local Policy Database, https://database .aceee.org/city/oakland-ca.

89. Nowak et al., *Leaders of the Pack: ACEEE's Third National Review of Exemplary Energy Efficiency Programs*, Report Number U132 (Washington, DC: ACEEE, 2013).

90. http://www.energy.ca.gov/renewables/.

91. http://www.energy.ca.gov/commission/fact_sheets/documents/core/EEAchieving _Energy_Efficiency.pdf.

9. Faster, Further, Farther: Empowering the Great Energy Transition

1. Felix Creutzig et al., "Towards Demand-Side Solutions for Mitigating Climate Change," *Nature Climate Change* 8 (2018): 260–271; Fergus Green and Richard Denniss, "Cutting with Both Arms of the Scissors: The Economic and Political Case for Restrictive Supply-Side Climate Policies," *Climatic Change* 150, nos. 1–2 (2018): 73–87; F. Geels et al., "Reducing Energy Demand Through Low Carbon Innovation: A Sociotechnical Transitions Perspective and Thirteen Research Debates," *Energy*

Research and Social Science 40 (2018): 23–35; F. Creutzig et al., "Beyond Technology: Demand-Side Solutions for Climate Change Mitigation," *Annual Review of Environment and Resources* 41 (2016): 173–198.

2. F. W. Geels, "Regime Resistance Against Low-Carbon Energy Transitions: Introducing Politics and Power in the Multi-Level Perspective," *Theory, Culture and Society* 31, no. 5 (2014): 21–40.

3. N. Melton, J. Axsen, and D. Sperling, "Moving Beyond Alternative Fuel Hype to Decarbonize Transportation," *Nature Energy* 16013 (2016).

4. P. J. Loftus et al., "A Critical Review of Global Decarbonisation Scenarios: What Do They Tell Us About Feasibility?" *Wiley Interdisciplinary Reviews (WIREs): Climate Change* 6, no. 1 (2015): 93–112.

5. C. Spreng, B. K. Sovacool, and D. Spreng, "All Hands on Deck: Polycentric Governance for Climate Change Insurance," *Climatic Change* 139, no. 2 (2016): 129–140.

6. N. Stern, *The Stern Review: Report on the Economics of Climate Change* (London: Cabinet Office—HM Treasury, 2006).

7. David L. Greene, "Uncertainty, Loss Aversion, and Markets for Energy Efficiency," *Energy Economics* 33, no. 4 (2011): 608–616.

8. Travis Bradford, "A Brief History of Energy," in *Solar Revolution: The Economic Transformation of the Global Energy Industry* (Cambridge, MA: MIT Press, 2006), 23–43.

9. International Renewable Energy Agency (IRENA), *Perspectives for the Energy Transition: Investment Needs for a Low-Carbon Energy System*, 2017, https://www.irena .org/publications/2017/Mar/Perspectives-for-the-energy-transition-Investment -needs-for-a-low-carbon-energy-system; C. Figueres et al., "Three Years to Safeguard Our Climate," *Nature* 546 (2017): 593–595.

10. D. Whetten and K. Cameron, *Developing Management Skills: Global Edition* (London: Pearson Higher Education, 2018).

11. https://www.bhg.com/home-improvement/advice/24-tips-for-energy-efficient -homes/.

12. https://www.directenergy.com/learning-center/energy-efficiency/25-energy-efficiency -tips.

13. https://www.thegreenage.co.uk/100-ways-to-save-energy-in-your-home/.

14. http://www.powerhousetv.com/Energy-EfficientLiving/Energy-savingsTips/027471.

15. https://www.ovoenergy.com/guides/energy-guides/120-ways-to-save-energy.html.

16. https://www.makeuseof.com/tag/7-energy-saving-technologies-lower-homes-carbon -footprint/.

17. https://www.moneycrashers.com/green-energy-technologies-solutions-home -improvement/.

18. https://www.powershop.com.au/23-gadgets-improve-energy-efficiency/.

19. https://www.geenergyconsulting.com/.

20. https://www.evainc.com/.

21. For example, in the United States, the Department of Energy offers a site that allows residential consumers to find consultants in their communities: https://www.energy .gov/energysaver/home-energy-audits/professional-home-energy-audits.

22. M. E. Porter and C. V. D. Linde, "Green and Competitive: Ending the Stalemate," *Harvard Business Review* 73, no. 5 (1995): 120–134.

23. P. Hawken, A. Lovins, and L. H. Lovins, *Natural Capitalism* (New York: Little, Brown, 1999).

24. S. V. Valentine, "The Green Onion: A Corporate Environmental Strategy Framework," *Corporate Social Responsibility and Environmental Management* 17, no. 5 (2010): 284–298.

25. S. V. Valentine, "Policies for Enhancing Corporate Environmental Management: A Framework and an Applied Example," *Business Strategy and the Environment* 21, no. 5 (2012): 338–350.

26. For more on these success stories, see https://www.theguardian.com/sustainable-business/environmentally-friendly-sustainable-business-profitable.

27. For more on this program, see http://files.ecomagination.com/wp-content/uploads/2010/07/Treasure_Hunt_FAQs_071110.pdf.

28. https://www.atkearney.com/consumer-goods/article?/a/the-profitable-shift-to-green-energy.

29. Wei Su, "Reducing Energy Use Is a Big Winner for Business and the Climate," *The Conversation*, May 25, 2016, accessed April 3, 2018, https://theconversation.com/reducing-energy-use-is-a-big-winner-for-business-and-the-climate-59659.

30. Su, 2016.

31. For more on this subject, see http://www.energyproductivity.net.au/.

32. K. S. Benjamin, *The Dirty Energy Dilemma: What's Blocking Clean Power in the United States* (Westport, CT: Praeger, 2008).

33. Benjamin, 2008.

34. Paul C. Stern et al., "Opportunities and Insights for Reducing Fossil Fuel Consumption by Households and Organizations," *Nature Energy* 16043, doi:10.1038/nenergy.2016.43.

35. For more on this project, see ABC News, "Elon Musk's Giant Lithium Ion Battery Completed by Tesla in SA'S Mid North," November 23, 2017, accessed July 17, 2018, http://www.abc.net.au/news/2017-11-23/worlds-most-powerful-lithium-ion-battery-finished-in-sa/9183868.

36. M. A. Brown, F. Southworth, and A. Sarzynski, *Shrinking the Carbon Footprint of Metropolitan America* (Washington, DC: Brookings Institution, 2008).

37. I. Kubiszewski et al., "Beyond GDP: Measuring and Achieving Global Genuine Progress," *Ecological Economics* 93 (2013): 57–68.

38. C. Hamilton, "The Genuine Progress Indicator Methodological Developments and Results from Australia," *Ecological Economics* 30, no. 1 (1999): 13–28.

39. A. Hsu, *Global Metrics for the Environment* (New Haven, CT: Yale University Press, 2016).

40. F. Caprotti, "Critical Research on Eco-Cities? A Walk Through the Sino-Singapore Tianjin Eco-City, China," *Cities* 36 (2014): 10–17.

41. B. K. Sovacool, M. A. Brown, and S. V. Valentine, *Fact and Fiction in Global Energy Policy: Fifteen Contentious Questions* (Baltimore, MD: Johns Hopkins University Press, 2016).

42. K. Finley, *Trust in the Sharing Economy: An Exploratory Study*, University of Warwick, 2013, https://warwick.ac.uk/fac/arts/theatre_s/cp/research/publications/madiss/ccps_a4_ma_gmc_kf_3.pdf.

43. National Academies of Sciences, Engineering, and Medicine, *Pathways to Urban Sustainability: Challenges and Opportunities for the United States* (Washington, DC: The National Academies Press, 2016), appendix A; J. Orsi, "The Sharing Economy Just Got Real," *Shareable*, September 16, 2013, accessed April 20, 2015, http://www.shareable.net/blog/the-sharingeconomy-just-got-real.

44. Sovacool, Brown, and Valentine, 2016.

45. Based on data published in S. Gupta et al., "Cross-cutting Investment and Finance Issues," in *Climate Change 2014: Mitigation of Climate Change. Contribution of Working Group III to the Fifth Assessment Report of the Intergovernmental Panel on Climate Change* (Cambridge: Cambridge University Press, 2014), 1211.

46. Marilyn A. Brown and Li Yufei, "Carbon Pricing and Energy Efficiency: Pathways to Deep Decarbonization of the US Electric Sector," *Energy Efficiency* 12, no. 2 (2019): 463–481.

47. Q. M. Liang and Y. M. Wei, "Distributional Impacts of Taxing Carbon in China: Results from the CEEPA Model," *Applied Energy* 92 (2012): 545–551, doi:10.1016/j.apenergy.2011.10.036.

48. This site is accessible at https://www.nyserda.ny.gov/About.

49. Krister P. Andersson and Elinor Ostrom, "Analyzing Decentralized Resource Regimes from a Polycentric Perspective," *Policy Sciences* 41 (2008): 71–93.

50. https://www.shortlist.com/entertainment/sport/george-foreman-on-ali/72991.

Index